# ZERO G

# ZERO G

## LIFE AND SURVIVAL IN SPACE

Peter Bond

CASSELL

This book is dedicated to my beloved wife, Edna

First published in the United Kingdom in 1999 by Cassell

Distributed in the United States of America by Sterling Publishing Co. Inc.
387 Park Avenue South, New York, NY 10016-8810

A CIP catalogue record for this book is available from the British Library

ISBN 0-304-35075-3

Designed by Les Dominey

Colour separation by Tenon & Polert Colour Scanning Ltd.

Printed and bound in China

Cassell
Illustrated Division
The Orion Publishing Group
Wellington House
125 Strand
London WC2R 0BB

*page 1*
*Buzz Aldrin admires the Stars and Stripes, a little crumpled but finally hanging freely, after a long struggle to deploy it.*

*page 2*
*Eagle's ascent stage closes in on the command module after the historic first lunar landing.*

# CONTENTS

# INTRODUCTION

When I was a child I would sometimes wonder what it would be like to fly like Peter Pan. I knew, however, that the experience of being able to turn somersaults in mid-air, to sleep and eat upside down and to float freely in any direction belongs in Never Never Land. Or does it?

Some 400 humans – astronauts, cosmonauts, call them what you will – have actually experienced these things and have entered another dimension in which the restrictions of gravity do not exist. Forced to adapt to an environment where there is no up or down, where an object has no weight and where they have the strength of super-beings, this rare breed has begun to open up a new frontier – the limitless bounds of outer space. This book is their story.

No words can adequately describe what it is really like to live 200 miles above Earth, shut inside a metal canister no larger than a mobile home for weeks or months at a time. The intention of this book is to give a flavour of the trials and tribulations, the dangers, the excitement and the monotony that are part of human space travel.

The illustrations are a vital part of the book. The inclusion of some of the most interesting and unusual photographs from the vast store of images gathered during several hundred missions will, I hope, enable all readers to appreciate the remarkable way of life of these latter-day Peter Pans.

Almost 40 years after Yuri Gagarin broke the gravity barrier and became the first person to travel in space there are still those who would turn back the clock and pretend that his flight and the hundreds since had never taken place. Money and resources, they argue, are wasted on sending humans to perform tasks that could just as easily be performed by robots.

If we follow this argument to its logical conclusion, we might as well scrap the shuttle fleet, refuse to build any more space stations and accept that we shall always be Earthbound, tied to a small blue planet that circles an ordinary star in an ordinary galaxy. Our exploration of the cosmos will always be at second-hand, dependent on a mass of micro-chips and electronic wizardry.

The alternative is to build on the human courage and achievements of the past four decades and to find new ways of exploring. If NASA and private enterprise can build faster, cheaper, better robotic spacecraft, it should not be beyond our capabilities to do the same for ships equipped to carry people. The challenge for the new millennium is to look forward to a time when space travel will be open to every citizen of planet Earth and when nations will work together to establish

*An IMAX (wide-screen) camera view of the shuttle* Atlantis *separating from the orange docking tunnel on the nose of the* Kristall *module. Also visible are the* Soyuz *ferry (top),* Spektr *module (right) and the* Kvant 2 *module (left).*

the first settlements on the Moon, Mars or some other far-flung outpost.

Many sources of information – too many to list here – have been drawn on during the preparation of this book. Special mention must be made of the wonderful encyclopedia of information provided by NASA Web pages and by the more traditional, but equally invaluable, printed pages of *Spaceflight* magazine. Particular thanks go to those who have responded to my interminable requests for photographs and other information, notably Debbie Dodds and Debra Herrin at Johnson Space Center; the Imaging Department staff at NASA HQ, the public relations staff at ESA, DLR and NASDA; the staff at CNES Diffusion; Kari Kelley at Boeing; and E. Rudolf van Beest and Chris Faranetta of Energia USA. Thanks also to my editor, Barry Holmes, whose humour and persistence eventually brought this book into existence, and to my wife, Edna, whose patience, understanding and innumerable cups of coffee enabled me to complete the endurance course.

*Peter Bond*

# 1

# SO YOU WANT TO BE AN ASTRONAUT?

More than 400 citizens of planet Earth have left their homes and families behind to venture into outer space. They were not supermen or women, but they had certain qualities that enabled them to achieve what most of us can only dream about.

## THE SELECT FEW

Finding the right people to explore the 'final frontier' proved to be a major headache. A number of occupational groups were thought to have ideal qualifications – submariners, for example, were used to long-term isolation, and mountaineers were accustomed to great heights and rarefied atmosphere – but eventually it was decided that military pilots would be the best choice. Their jobs required a familiarity with modern technology and high altitude flight, and every day they were expected to display rapid reactions, to cope with stressful situations and to make instant, life-or-death decisions.

So it was that the first astronauts were experienced military fliers, mostly drawn from the ranks of the test pilots. The Mercury Seven consisted of three Navy pilots, three from the Air Force and one Marine. At the time of their selection in 1959, their ages ranged from 32 (L. Gordon Cooper) to 37 (John Glenn). They stood between 5 feet 7 inches (Virgil Grissom) and the maximum allowed, 5 feet 11 inches (Alan Shepard), and hit the scales at between 150 pounds (Cooper) and 185 pounds. (Walter Schirra had to lose 5 pounds to achieve the 180-pound limit.)

Several of these men were war heroes. Glenn had received the Distinguished Flying Cross (DFC) and Air Medal for service in the Second World War and the Korean War. Grissom and Schirra had also received the DFC and Air Medal in the Korean War.

*America's first astronauts, the Mercury Seven, were all experienced military pilots. From left to right: M. Scott Carpenter, L. Gordon Cooper, John Glenn, Virgil Grissom, Walter Schirra, Alan Shepard and Donald ('Deke') Slayton.*

Not everyone was impressed by the career of 'astronaut' or 'star traveller'. Some flyers, such as world speed record holder Chuck Yeager, dismissed the Mercury programme as 'man-in-a-can' and chuckled when chimpanzees preceded humans on the preliminary flights. If a chimp could handle a capsule, why bother to send up a qualified pilot?

In the Soviet Union the emphasis was more on youth and fitness than on experience. Yuri Gagarin, who was 27 years old when he made his historic orbit of Earth, had clocked up just 230 hours in the air when he was chosen in 1959. His back-up, Gherman Titov, was still only 25 years old when he orbited Earth in 1962. There were exceptions. Sergei Korolev, chief Soviet designer, persuaded 43-year-old Georgi Beregovoi to join

The Vostok and Voskhod cosmonauts pose together in 1965. Back row (from left to right): Alexei Leonov, Gherman Titov, Valeri Bykovsky, Boris Yegorov and Pavel Popovich. Front row (from left to right): Vladimir Komarov, Yuri Gagarin, Valentina Tereshkova, Andrian Nikolayev, Konstantin Feoktistov and Pavel Belyayev.

the ranks of the cosmonauts in 1964. The war veteran was already a Hero of the Soviet Union and an experienced test pilot, having flown more than 4000 test flights and clocked up over 2500 hours of flying time.

For the Soviets selection was only the first step on the road to fame. The names of the cosmonauts were not released to the public until they had flown a mission, and when one of the initial selection, Valentin Bondarenko, was burned to death while training in the oxygen-rich atmosphere of an isolation chamber, his death was covered up for 25 years. Another three of the original 20 cosmonauts were dismissed from the corps after a skirmish with a military patrol, but, again, the incident was covered up and their identities remained secret. Indeed, eight of the original group remained Earthbound and anonymous until their names leaked out decades later.

Ambition, sound education and hard work enabled most of the Soviet pioneers to overcome modest upbringings in postwar USSR. Gagarin grew up in a remote village, the son of a collective farmer and a dairy maid. Titov was something of an exception – he was the son of a language teacher, who was also a mechanic and taxi driver.

After Gagarin's flight in 1961, Korolev decided it was time to recruit some engineers and scientists, and despite the opposition of the Ministry of Defence, 13 engineers were eventually passed for flight training in April 1964. The first result of this change in direction came later that year when one of Korolev's young protégés, Konstantin Feoktistov, and physician Boris Yegorov were assigned to *Voskhod 1*.

Selection procedures gradually changed in the United States too. A 1962 recruit named Neil Armstrong, a veteran of the Korean War and a former X-15 rocket plane pilot, became the first civilian astronaut to fly in space. Then, in 1965 and 1967, the National Aeronautics and Space Administration (NASA) chose its first civilian scientists in preparation for the Moon landings and the proposed space station. Although only geologist Harrison Schmitt set foot on the Moon, several others flew on board *Skylab*, and physician Story Musgrave went on to become a six-time astronaut.

The armed forces did not relinquish their grip entirely, however. Military officers were recruited for Pentagon space projects in the 1960s. The DynaSoar and Manned Orbiting Laboratory (MOL) programmes were cancelled, but a number of MOL candidates moved over to NASA and eventually flew on the shuttle. Other Department of Defense (DOD) astronauts were later selected to participate in secret shuttle missions. They were kept away from the limelight and barred from discussing their training or orbital activities. Most current shuttle pilots have a military background.

## THE OBSTACLE COURSE

Would-be astronauts from both the USA and the then Soviet Union had to undergo an exhausting and exhaustive series of tests to reach their goal. Many dropped by the wayside. Some, such as Charles (Pete) Conrad, failed the first time, only to finish the race the second time around as a member of the 1962 astronaut group. Conrad eventually flew on four missions and became one of only 12 men to walk on the Moon.

Those who completed the course often complained about the invasiveness of the procedures. Test pilots familiar with the forces generated by rapid spins and supersonic skydives found the physiological trials the easiest to overcome. Rotating around a giant room at the end of a centrifuge until the occupant blacked out at around 9G was considered a piece of cake.

Experienced test pilots similarly dismissed the examinations involving reactions to extreme heat and cold. 'Your face gets kind of warm in that oven, and I had a couple of hot spots on my knees and hands. It was uncomfortable, but you can tolerate it. It was also a pretty good jolt when they stuck our feet into a bucket of ice cubes. But before the 7 minutes were up my feet got so cold I couldn't even feel them, so that was not so bad,' said Donald ('Deke') Slayton.

The candidates were also monitored while running on a treadmill and doing the Harvard step test. Even hard-bitten fliers like Slayton found this a challenge. 'You jump up and down off the platform for about 5 minutes. Then they throw you on the tilt board, which stands you straight up while they measure your blood pressure and your heart rate.'

Occasionally, even passing the selection course was not enough. After Slayton became one of the Mercury team, he was barred from flying on medical grounds. A heart murmur detected on one of his centrifuge runs led to his being grounded just 10 weeks before he was scheduled to lift off on America's second orbital mission. Scott Carpenter moved up the flight list while Slayton was declared unfit for operational duty and, perhaps most hurtful of all, given a life ban from flying solo.

For once, the hard luck story had a happy ending. Despite accepting a desk job as Director of Flight Crew Operations, Slayton continued to fly high-performance aircraft in the hope that one day the authorities would realize their mistake. Then, some 10 years after his ambitions had been shattered, a NASA review board gave him a clean bill of health. At the age of 51 years, the veteran participated in the US-Soviet Apollo-Soyuz Test Project.

Disqualification on medical grounds was fairly common during the early missions. Amazingly, some of the first pioneers were fortunate enough to escape the doctors' eagle eye or to persuade them to look the other way. Georgi Grechko benefited twice from such generosity. A hand injury suffered during the Second World War was ignored, while a broken leg incurred during training delayed his first flight by several years but did not disqualify him. Another early selection, Titov, managed to hide an old wrist fracture and became the second Soviet man in space.

Occasionally, a medical emergency has had a profound influence on events. *Apollo 8* command module pilot Michael Collins had a slipped disc in his neck, which was pressing on his spinal cord. Aware he could never fly again in this condition, Collins opted for two dangerous operations. His courage paid off. Instead of participating in *Apollo 8*, he is remembered as a member of the first Moon landing mission.

*Shuttle crews include different types of astronaut. In this portrait of the STS-87 crew are (back row) mission specialists Winston Scott (left) and Takao Doi from Japan; and (front row, from left to right) mission specialist Kalpana Chawla, pilot Steven Lindsey, commander Kevin Kregel and Ukrainian payload specialist Leonid Kadenyuk.*

## So You Want to be an Astronaut?

NASA constantly needs to replace astronauts who retire. Applications from any US citizen who wants to participate in the shuttle and International Space Station (ISS) programmes are accepted, with a biennial selection of astronaut candidates. A stringent sifting process means that only a few have a realistic chance of crossing the final tape.

There are currently three classes of full-time astronaut. Pilot astronauts are responsible for controlling and operating the orbiter. After one or more flights in the right-hand seat, pilots may be given their own command, with overall responsibility for the success of the mission and the safety of the $2-billion vehicle and crew. The pilot and commander may also be called upon to help with the robot arm, extra-vehicular activity (EVA) or other payload operations. During the first four test flights of the shuttle, the pilot and commander were the only crew on board.

Anyone hoping to pilot a shuttle must meet certain basic requirements:
❶ a bachelor's degree (or preferably, an advanced degree) from an accredited institution in engineering, biological science, physical science or mathematics;
❷ at least 1000 hours pilot-in-command time in jet aircraft (test pilot experience is highly desirable);
❸ an ability to pass a NASA Class 1 space physical examination, which includes eyesight 20/70 or better uncorrected, correctable to 20/20 in each eye and blood pressure of 140/90 measured in a sitting position;
❹ height between 5 feet 4 inches and 6 feet 4 inches.

Mission specialists are assigned specific tasks related to operation of the payloads, crew activities, consumption of power and other consumables, and space station assembly and operations. Jobs involving scientific experiments, use of the remote arm or spacewalks are the domain of the mission specialist. Candidates must meet the following requirements:
❶ a bachelor's degree from an accredited institution in engineering, biological science, physical science or mathematics, followed by an advanced degree and/or up to three years of 'related, progressively responsible, professional experience';
❷ ability to pass a NASA Class 2 space physical, which include: eyesight 20/200 or better uncorrected, correctable to 20/20 in each eye and blood pressure of 140/90 measured in a sitting position;
❸ height between 4 feet 10¹⁄₂ inches and 6 feet 4 inches.

Even if you meet these requirements and your references are accepted, there is then the matter of a week-long series of interviews, medical evaluations and orientation. Complete background investigations are carried out on those finally selected.

Once chosen, astronaut candidates are assigned to Johnson Space Center (JSC) in Houston, Texas, for a one- to two-year period of training and evaluation. If they complete the course satisfactorily, civilians are offered permanent positions as employees of the US government and will be expected to remain with NASA for at least five years.

Procedures are slightly different for military personnel. The initial sifting of applications is carried out by the respective military service. New recruits are detailed to JSC but remain on active-duty status for pay, benefits, leave and other service matters.

The third class of astronaut is the payload specialist, the first of whom flew on the Spacelab in December 1983. These specialists are chosen by the company or organization that built or sponsored the particular payload. Unlike the professional astronauts employed by NASA, payload specialists are scientists, engineers or physicians assigned for their knowledge of specific satellites or experiments. A number of payload specialists have not been US citizens, starting with German physicist Ulf Merbold on the maiden flight of Europe's Spacelab (STS-9).

In the more relaxed days before the loss of *Challenger* a number of unusual space travellers were able to hitch a ride. Among them were Senator Jake Garn and Congressman William Nelson, two politicians with responsibilities for overseeing the space programme, who justified their selection by acting as human guinea pigs in studies of motion sickness.

Ironically, perhaps the most famous representative of the non-professional astronauts never experienced what it was like to see our planet from orbit. Christa McAuliffe, then 37 years old, was chosen from 11,000 applicants to become the first schoolteacher in space. She cheerfully admitted that she could not really be described as an astronaut, declaring a preference for the term 'space participant'. Sadly, she was never able to describe the wonders of the cosmos to her pupils.

The disaster of 28 January 1986, when *Challenger* was destroyed, underlined the message that spaceflight is dangerous and unpredictable. Outsiders were not allowed to fly on the shuttle for another 10 missions, until December 1990. A NASA proposal for a competition to find a space journalist was abruptly dropped.

This attitude changed in January 1998 when NASA announced that it had selected Barbara Morgan, a schoolteacher from McCall, Idaho, to join the next astronaut candidate class as a mission specialist. From now on, mission specialists with backgrounds in teaching science, mathematics and technology will be open to selection as astronauts. Their task will be to carry out educational programmes in addition to their other assigned flight duties.

## BREAKING THE BARRIERS

Today, there are almost no limits on the age, gender or skin colour of space travellers. It was not always so. While the Mercury astronauts tended to be experienced test pilots and more mature in years (Alan Shepard was 37 years old when he flew into space, John Glenn was 40 years old), their Soviet counterparts were mostly in their mid-20s. Since then, with the introduction of the shuttle and a consequent reduction in levels of bodily stress, the upper age limit has been continually climbing. The oldest person to fly in space is Glenn, who made his second spaceflight at the age of 77. The oldest spacewoman, Shannon Lucid, a relative youngster at 53, was also American.

Opportunities for non-whites were also at a premium for a long time. The breakthrough was made by Cuban Air Force Colonel Arnaldo Tamayo Mendez when he flew as a guest aboard the *Salyut 6* space station in September 1980. From selling vegetables and shining shoes when he was 13 years old, Tamayo Mendez worked his way through college and military academy until he achieved command of a Cuban air unit before spending almost 8 days in space.

Not until August 1983 did the first African-American astronaut fly on a US mission. Mission specialist Lieutenant Colonel Guion Bluford was among those selected by NASA in 1978. During the 6-day mission he was given responsibility for a cell experiment designed to develop a commercial treatment for human diabetes.

If it took a while to break down the age and race barrier, the gender barrier was breached at an early stage and then re-erected. The woman who broke the male monopoly

The first African-American astronaut, mission specialist Guion Bluford, tries out the exercise treadmill on Challenger's mid-deck during STS-8.

was a 26-year-old factory worker, Valentina Tereshkova, who became the sixth Russian in space on 16 June 1963. While the USA was struggling to keep its Mercury men aloft for more than 36 hours, Tereshkova spent 48 orbits and almost 71 hours in space, longer than all the astronauts together.

Not surprisingly, there were those in the West who belittled Tereshkova's feat, calling it a propaganda stunt, but the Soviet heroine emphatically dismisses such comments as sour grapes:

*Preparation for the spaceflight took nearly two years' training. We started at the end of 1961 and were only ready in 1963. We were not spared any section of training. We were not given an easier time just because we were women.*

Simply being chosen out of 400 applicants was a feat in itself. Advertisements for parachutists and fliers who could fill the female space slot were posted in air clubs around the country. Candidates were interviewed by the deputy commander-in-chief of the Air Force and there were intensive medical checks. Five women survived the process.

Inexperienced they may have been (although Tereshkova, for example, had completed 126 parachute jumps), but from 14 March 1962 they had to undergo the same training as their male colleagues: centrifuge, isolation in a vacuum chamber, zero gravity trips in aircraft and parachute jumps wearing a space suit. The academic preparation included spacecraft construction, navigation and geophysics.

Tereshkova established another milestone on the road to gender equality when she married cosmonaut Andrian Nikolayev. Suggestions that the demanding training and exposure to space radiation could damage a woman's ability to conceive were refuted in June 1964 when she bore a healthy daughter, 7 months into her pregnancy. She attributed the premature birth to her exhausting post-flight schedule of overseas trips, academic studies and housekeeping.

Following Tereshkova's debut, there were hopes of a two-woman Voskhod flight

US Air Force Lieutenant Colonel Eileen Collins, the first woman to command a space mission. Collins was selected by NASA in 1990. She flew as pilot on the STS-63 Mir rendezvous mission in February 1995 and revisited Mir in May 1997. Her chance to occupy the commander's seat will come in summer 1999 in a mission to deploy the Chandra X-ray telescope.

that would include an EVA. This never took place, and the women's group was disbanded in 1969. Eleven years passed before women were again selected for the Soviet space programme.

It is curious that, although the Soviet Union placed a woman in orbit as early as June 1963, 20 years before Sally Ride broke the gender barrier in the United States, only two other Russian women have experienced zero gravity. Test pilot Svetlana Savitskaya twice visited the *Salyut 7* space station in the early 1980s and was allowed to carry out a spacewalk. Engineer Elena Kondakova, the wife of cosmonaut and flight director Valeri Ryumin, spent 170 days on *Mir* in 1995 and later flew on the shuttle.

The only other women to fly on Russian craft were British chemist Helen Sharman, French medical doctor Claudie André-Deshays and American biochemist Shannon Lucid, who broke the female endurance record with 188 days aboard *Mir* in 1996.

The arrival of women in the macho world of test pilots and astronauts required varying degrees of adjustment by the male community. Despite more overtly sexist attitudes on the Russian side, Tereshkova was full of praise for her male colleagues:

*Their attitude towards us was fair and helpful. For instance, if we needed extra training they were ready to spare the time and energy in the evenings, in the library, or for other training facilities like the simulators. Not once did any of the men say they would not find time to help us.*

Not all women have been treated as gallantly by their male counterparts. Helen Sharman recounts how, before leaving for the launch pad, Alexei Leonov gave her a pink chiffon jump-suit with a frilly front and billowing sleeves, explaining: 'I thought you might like to dress for dinner.'

Savitskaya received similar treatment. Valentin Lebedev recounted how they 'saw her sitting in the docking compartment combing her hair' before she entered the *Salyut 7* space station. Once she was on board, the men presented her with a bunch of flowers, but they could not resist a joke at her expense before the first meal by giving her a blue floral-print apron.

Today, NASA has an 'affirmative action' programme to encourage qualified minorities and women to apply as astronaut candidates. Most crews contain women, and female pilots are now accepted as part of the normal scheme of things. In mid-1999 Eileen Collins will become the first woman to command a space mission.

It was not always so. Until the shuttle arrived on the scene, no women were even considered for inclusion in the manned programme, and not until 18 June 1983 did Sally Ride become America's first woman astronaut. One of six women selected by NASA as mission specialists in 1978, Ride struggled to rise above the media circus that surrounded her flight. Every moment of her life was scrutinized in excruciating detail by the media, and pre-flight interviews plumbed new depths of triviality, many of them focusing on how she would cope with being in a confined space with four men.

After the return to Earth, she faced a deluge of questions from reporters and offers from Hollywood agents. Maintaining her dignity, she rejected their advances: 'I didn't go into the space programme to make money or be famous.'

## INTERCOSMOS

For many years the only way to fly on a Soviet rocket was to be a Soviet citizen. This changed on 16 July 1976 when the member states of the Intercosmos organization – the USSR and eight other countries – signed an agreement that opened the way for outsiders to experience at first hand the benefits of the communist system; Vietnam joined the organization in 1979. The first group of six foreign cosmonauts began training at Star City near Moscow in December 1976. Eventually, nine flights involving Intercosmos members took place between 1978 and 1981.

Most of these missions followed the same format. Each country nominated two candidates, one of whom was selected after two and a half years of training. The chosen one would be given an assignment to photograph his homeland and carry out some home-grown experiments, which were usually related to biology and space medicine. The trip into space would last about 8 days and include a stay on board the *Salyut 6* space station. The training included housekeeping, emergency procedures and learning Russian, but the sole responsibility for the technical aspects of the mission remained in the hands of the Russian commander.

Although such trips were certainly of propaganda – and, later, monetary – value to the Soviets, how much the sponsor nations got out of them is debatable. On several occasions, the flights almost ended in tragedy. In the case of the failed *Soyuz 33* mission involving Bulgarian cosmonaut, Georgi Ivanov, the Soviets provided compensation nine years later in the form of a trip to *Mir* by his compatriot Alexander Alexandrov.

All nine Intercosmos missions used the old two-man Soyuz, but the introduction of an improved Soyuz-T spacecraft in 1980 allowed the Soviets to offer a third seat to a paying customer. The first to take advantage of this cut-price space trip was French 'spationaut', Jean-Loup Chrétien, who visited the *Salyut 7* space station in 1982. The importance the authorities attached to this first flight by a Westerner on a Soviet spacecraft was illustrated by the blanket coverage it received in the communist media.

Since then it has become common for foreigners to pay for short visits to Soviet/Russian space stations. The list includes representatives of France, Germany, Austria, Britain, Japan, India, Afghanistan and Syria. The most unusual of these brief trips were those by Briton Helen Sharman and Japanese TV reporter Toyohiro Akiyama. Sharman's 1991 flight was almost cancelled when the promised commercial backing failed to materialize, resulting in a significant financial loss for the Soviets. On the other hand, Tokyo Broadcasting Service was happy to pay $25 million for Akiyama's week of weightlessness in December 1990.

The shuttle programme has also had its share of foreign participants. Most of these have been Japanese and Europeans, who were usually participating in Spacelab experiments, or Russians involved with the shuttle–*Mir* missions. The highest ranking foreign dignitary to have flown on a shuttle was Prince Sultan Salman Abdul Aziz, 28-year-old nephew of King Fahd of Saudi Arabia. The prince went along to oversee the launch of the Arabsat telecommunications satellite in his capacity as Acting Director of the Saudi TV Commercial Department.

## OTHERS TAKE UP THE CHALLENGE

Although the two space super-powers have provided the majority of the world's space explorers, a number of other countries are keen to build a home-grown form of human space transportation for their astronauts.

Many of the 14 member states of the European Space Agency (ESA) have already seen one or more of their citizens take a ride on either the shuttle or a Soyuz. German Thomas Reiter led the way with a 6-month sojourn on the *Mir* station. French astronauts have visited Russian space stations on six occasions since 1982 as well spending time on board the shuttle.

Despite financial problems during the 1990s, most of the member states intend to participate in the International Space Station (ISS), carrying out research based on the European Columbus laboratory. In March 1998 the ESA announced that all astronauts from member states would be reassigned to a single corps to prepare for the ISS. By the year 2000 this group will consist of 16 astronauts based in the European Astronaut Centre in Cologne, Germany. Europe's ambitions are open for all to see, but much less is known about the plans of the Chinese, who seem intent on continuing the policy of secrecy that dogged the Soviet space programme for so long. However, with one-fifth of the world's population and a rapidly developing economy, China has made it clear that space technology will play a major role in establishing the country as the next superpower.

Over the years China has several times announced its intention to send men into orbit in the not-too-distant future, but details have generally been sketchy. As early as 1979–80, a number of photographs were published that seemed to show Chinese astronauts in training, but then all went quiet until 1986, when officials again admitted their interest in a manned programme. A report in the *People's Daily* declared: 'We have already succeeded in producing life-support systems and in solving the problems of controlling gas composition and pressure in the cabin and the level of heat and humidity.'

After another hiatus, the National Space Administration announced that it would present a plan to the government in 1995 calling for the construction of a two-man capsule that could be launched by the year 2000. The spacecraft would be based on existing Chinese unmanned recoverable capsules. Plans called for a new launch pad and facilities for manned flights to be built in the Gobi Desert near the Russian border. Later that year it was announced that China wanted to learn more about Russian life-support systems and cosmonaut training methods.

In 1996 two Chinese astronaut candidates were sent to Star City to undergo a tough training course. Two years later Ma Xingriu, vice-president of the Chinese Academy of Space Technology, told an international conference: 'China is striving to make breakthroughs in manned spaceflight technology at the end of this century or the beginning of the next century.' Meanwhile, the authorities have admitted their intention to upgrade the Long March rockets so that they can launch a 20-ton payload, such as a space station, into low Earth orbit.

## TEAM EFFORT

Becoming a fully fledged astronaut with your name on the active list is only the first step on the road to the stars. The most frustrating part follows – waiting for an assignment. Some qualified astronauts have had to wait many years for a mission; others have never made it into orbit at all.

Once a group of astronauts has been assigned to a mission, perhaps a year before the scheduled lift-off date, it is the responsibility of the commander and the ground support staff to mould the disparate group into an efficient crew capable of dealing with all eventualities. As the individuals get to know each other and work together during training, the team spirit gradually evolves.

Only occasionally does one person fail to blend with the rest of the team. In September 1981, for example, Yuri Malyshev was named as commander of the first joint-Soviet-French mission. The following February he was replaced, ostensibly on medical grounds. The French later revealed that the reason was a personality clash between Malyshev and the French prime candidate, Jean-Loup Chrétien.

Sometimes the process is disrupted when an astronaut or an entire crew has to be replaced at short notice, usually for medical reasons. The change that had the most disastrous consequences involved the crew of *Soyuz 11*, the first mission to occupy a space station. A few weeks before launch, Valeri Kubasov became ill and was replaced by his back-up, Vladislav Volkov. There followed a debate over whether to match Volkov with the other members of the back-up team. It was eventually decided to swap everyone. Instead of the prime crew, therefore, Volkov, Dobrovolski and Patsayev were the three to perish when the air leaked from their capsule during the return to Earth.

## SIMULATING ZERO G

Much crew training is just hard graft in the classroom: learning the spacecraft systems, learning how to stay alive and learning how to carry out the flight programme. For foreign astronauts unfamiliar with the language, this becomes doubly difficult. A number of astronauts training in Star City have commented that the worst part of their training was learning Russian. But probably the most difficult aspect is preparing crews for the unfamiliar – how to live and work in zero gravity while living on a one-gravity planet.

Simulators have existed since the late 1950s, but today they are more sophisticated and play an increasing part in training. They include highly realistic computer visual effects and motion to give as accurate an impression as possible of a ride into space. They allow the crew to go through a range of ascent and orbital entry scenarios, when flight controllers throw launch abort experiences and other problems their way. Sometimes they may have to deal with more than 200 'malfunctions' in one 4-hour simulator session. During this process, individual astronaut's weaknesses can be worked on and improved. All this helps to create the team spirit that is so essential for a successful mission.

Computer technology is playing an increasing role in the latest simulation techniques in the form of virtual reality, which enables astronauts to visualize various situations in three dimensions.

There is, however, one experience that not even the most powerful computers can simulate – that of floating weightless in space. One of the most common, although not always the most popular, ways of reproducing zero G has been to fly parabolas inside a large, stripped-down aircraft, the equivalent of travelling on a giant switchback in a fairground but without a seat belt. The Russians use an Ilyushin 76; the Americans use a KC-135; the French use a Caravelle.

As the aircraft levels out at the top of its climb, then pitches down as it accelerates towards the ground, its occupants experience perhaps 40 seconds of true weightlessness. When this is repeated numerous times in a session, astronauts familiarize themselves with the unique problems associated with donning pressure suits or carrying out a space-walk. The down-side is the effect this switchback ride has on all but the toughest constitutions – the early astronauts dubbed the aircraft the 'Vomit Comet'.

More complex tasks, such as simulating repairs to a space station or satellite, are usually undertaken in giant water tanks. These include the 40-foot deep neutral buoyancy simulator at Marshall Space Flight Center in Huntsville, Texas, and the 25-foot deep weightless environment training facility at JSC. Largest of all is the recently completed neutral buoyancy laboratory at Johnson Space Center (JSC), which is three times the size of the other tanks and is able to accommodate two simultaneous practice sessions. The Russians also have full-scale mock-ups of a Soyuz craft and the *Mir* space station in a 40-foot tank at Star City.

Assisted by divers using scuba equipment, the astronauts wear pressure suits, which are weighted so that they neither rise nor sink. Astronauts preparing for spacewalks spend many hours under water in one of these giant pools.

Alongside this ground-based training, astronauts are expected to hone their flying skills by practice in NASA's T-38 fighter planes. These aircraft are also available for visiting contractors and travel between NASA centres. In addition, shuttle pilots take to the air again and again in a specially modified Grumman aircraft, which simulates the orbiter's descent from 35,000 feet to landing. Even after each pilot astronaut has made 500 practice landings in this aircraft, landing the actual shuttle is a challenging task.

## SURVIVAL TRAINING

Modern training procedures are a far cry from the days of Mercury and Gemini. However, there are still a few nasty surprises along the way. Astronaut candidates are required to complete a military water survival programme before they begin their flight training. This includes treading water for 10 minutes, and swimming three non-stop lengths of a 27-yard pool, followed by three lengths wearing a flight suit and tennis shoes. In addition, they have to qualify as scuba divers before they can begin EVA training.

The countries and oceans over which astronauts fly vary according to the mission. Most early US flights were restricted to an inclination of 30 degrees, so they passed only over areas lying between 30 degrees north or south of the equator, which included all the tropics and extended as far north as the launch centre at Cape Canaveral in Florida.

More recently shuttle missions have flown as far as 57 degrees from the equator, which carries them over most of western Europe, the densely populated regions of Russia as well as all of the southern land-masses except Antarctica. All Russian manned missions fly at inclinations of 51.6 degrees.

When the crews are flying so far from the equator and relying on a ballistic re-entry in a Gemini, Apollo or Soyuz capsule, it is prudent for them to be trained to survive emergency touchdowns in all possible environments. The first astronauts underwent an imaginative series of tests. They were given a glimpse of jungle survival techniques in the Panama Canal Zone, which included sampling jungle cuisine. Jim Irwin described a memorable gourmet meal:

*The appetizer was braised boa, our entrées were iguana thermidor, oven-roasted wild boar and baked armadillo. We had heart of palm and bamboo shoot as the salad, and our dessert was tropical fruit compote.*

After a one-day introductory course, the men were airlifted into the rain forest wearing flying suits, long underwear and tennis shoes and left to their own devices. The next three days turned into an endurance test. They were soaked by tropical rains, came face-to-face with a large leopard (both participants ran in the opposite direction as fast as they could) and ended up eating heart of palm for breakfast, lunch and dinner. When the endurance test came to an end, they floated downstream on inflatable life-rafts to an Indian village where they dined on roast iguana.

Desert survival was not much more appetizing. The men were expected to construct makeshift tents and clothing from parachutes and to live off the meagre resources offered by scarce plants and wildlife.

Surviving an unplanned splashdown is an important skill for participants in both the American and Russian programmes. In the days of the Soviet Union, cosmonauts underwent water training exercises at Sochi in the Crimea. Today, Sochi is part of the Ukraine, so the training now takes place at Dujbga, 60 miles to the west. The training lasts about one week. Once on board, the mock-up of a Soyuz capsule, complete with safety collar, is lowered from the training ship and released at the end of a line. Seated alongside two trainers, each trainee takes it in turns to endure the claustrophobic, nausea-inducing sessions.

*The Mercury Seven looking distinctly unhappy and rather the worse for wear after completing four days of survival training in the Nevada Desert. From left to right: Gordon Cooper, Scott Carpenter, John Glenn, Alan Shepard, Virgil Grissom, Walter Schirra and Donald Slayton.*

Inside the closed chamber, the would-be cosmonaut has a choice of scenarios to simulate. For a high-latitude re-entry, he has to take off the pressure suit and clamber into four layers of Arctic survival clothing, topped by an orange dry-suit. For warmer waters, the routine is less arduous: simply put on a life-jacket and locate the survival kit.

After 1–2 hours of bobbing around in the stifling cabin, trying not to be seasick, the hatch is thrown open and the cosmonaut is able to demonstrate his ability to fall backwards into the sea. Pushing with the legs is not recommended because the capsule floats at an angle and is likely to ship water at the slightest excuse, resulting in the drowning of the other two occupants. Once in the water, the crews practise using the flares that would be used to attract a rescue plane, helicopter or ship. A 'pistolette' is provided to frighten away over-friendly sharks.

## PERSONAL PRESSURES

Once an astronaut has been assigned to a mission, the pressures of work tend to build. Astronauts and their families have always had to put up with long hours, prolonged absences from home and living in the public eye. This was particularly true in the 1960s, when there were fewer trainees and each mission received blanket media coverage. Family breakups and divorces became common.

After his retirement from NASA, Frank Borman, who flew on *Gemini 7* and *Apollo 8*, described how astronauts' domestic lives 'became a shambles'.

*While training for Gemini and Apollo, I averaged 250 days a year away from home. Jim Lovell's family used to call him 'the travelling salesman,' and I was just as bad. In fact, while I was preparing for Gemini 7, I could find every switch, knob and lever in the spacecraft blindfolded. But when I got home, I didn't know where Susan kept the water glasses.*

*We were never counselled on this difficult adjustment. It wasn't really NASA's fault, because the space agency itself wasn't prepared. And some of our adjustment problems were our own fault. Basically most of us still had the fighter pilot's macho hang-ups and would have regarded any psychological assistance as a form of weakness. So we expected our families to suppress their fears, look happy every time a camera was pointed in their direction and perform like trained seals. And what made it worse was that too many people in NASA expected this of them, too.*

For men like Borman, who had dedicated their lives to the pursuit of individual excellence on some lonely air base, life in a fishbowl was an alien existence. Instead of brief, curt exchanges over the intercom, they had to be prepared to wax lyrical about their exploits, becoming poets instead of pilots, actors instead of engineers.

Eventually, the men realized that it was time to move on to pastures new. Schirra commented: 'I quit while I was ahead and I'm glad I did .... As the space programme matured, so did I. I matured so much that sometimes I was afraid I was losing my sense of humour.'

For some, it was all too much. Buzz Aldrin became an alcoholic and underwent bouts of depression, which required psychiatric help and drug therapy before he was able to put together the pieces of his life.

## RISKS AND REWARDS

Risk is not taken into account in the determination of astronaut remuneration. In August 1997 salaries for NASA astronauts ranged from $48,000 to $103,000 a year, depending

*Fame and fortune? Alan Shepard proudly wears the Distinguished Service Medal as he waves to the crowds outside the Capitol building in Washington, D.C. Beside him is his wife Louise. Earlier he had spoken to Congress about his historic flight aboard Freedom 7.*

on years of government service and past promotions. As government employees, astronauts are not allowed to benefit financially from any sponsorship deals or contracts.

Things were rather different in the early days when the astronauts were few in number but continually bombarded by demands for personal appearances and interviews. NASA officials came up with a profitable, if controversial, way for the men to limit media attention and reduce the pressures – financial and personal – on them and their families. In return for awarding exclusive rights to their 'personal stories', the Mercury Seven signed a contract with *Life* magazine. At a time when it was impossible to buy additional life insurance because their jobs were considered too risky, $71,000 a year on top of their military pay allowed them to raise their standards of living and achieve considerable financial security while restricting media harassment. This bonus enabled them to build motels and invest in businesses that set them up for the rest of their lives.

As the number of astronauts increased, the financial rewards from this arrangement inevitably declined. In 1963 a new contract was agreed with *Life* and the Field newspaper syndicate for just over $1 million, but this had to be divided among 29 astronauts rather than seven – the equivalent of $34,500 a man. When Field dropped out of the deal in 1967, the pot was reduced to $200,000 a year. *Life* renegotiated the contract, so that each astronaut earned a further $5000 a year until the magazine folded in 1970.

There were other worthwhile perks associated with the job. The Mercury men, for example, were offered free homes near the Cape Canaveral launch centre, but as time went by the benefits were scaled down, although 'good deals' on mortgage rates remained commonplace.

'There were certain advantages in being an astronaut that I really enjoyed,' wrote *Apollo 15* astronaut Jim Irwin:

*First, General Motors offered us a car on a lease option for just token money. For a couple of years I had a red Corvette .... Then Ford joined the action, so I got a red Mustang convertible. Then the crews started getting cars that looked alike so they could be recognized as they travelled back and forth from the Cape. Apollo 12 had gold and black. Dave, Al and I decided that we would be the All-American team and go red, white and blue. ... As we moved closer to flight, we got later models and we had them put racing stripes on our cars. I had a red Corvette with a blue and white racing stripe. Mary was very embarrassed to have those two brand-new red cars in the garage.*

Unofficial fund-raising activities sometimes got the men into trouble. Astronauts often carried into space small items that they could later present to friends and family or sell as unique mementoes. The crew of *Apollo 15* took 650 envelopes to the Moon, 400 of which were unauthorized and smuggled onto the spacecraft inside David Scott's pressure suit. These envelopes were cancelled by the astronauts in their lunar 'post office'. Each crew member was to keep 100 of the extra envelopes, with the remainder going to a

German acquaintance, Horst Eiermann. The men expected to get at least $8000 for their allocation, although all agreed not to sell their envelopes before the Apollo programme was concluded.

The affair came to light when Eiermann's envelopes began to appear on the market. Individual items were selling for as much as $1500 each, and NASA eventually decided to act. The astronauts received only the second official reprimand in the agency's history. During testimony before the Senate Space Committee, the disgraced crew revealed all of the unauthorized articles they had secreted aboard the spacecraft. Scott and Worden were removed from the *Apollo 17* back-up crew and never flew in space again, and Irwin decided it was time to retire.

In Russia the system of remuneration is entirely different and working out future earnings is more complex. Cosmonauts receive an additional monthly living-in-space allowance as well as bonuses for high-risk activities such as EVA ($1000 extra) and manual docking ($1000 extra). The average income while in orbit is about $3000 a month, but a typical *Mir* crew might receive as much again in bonuses, bringing their total earnings to $40,000 for a 6-month mission. This may not sound generous, but the national average wage is $200 a month and the crews are paid in dollars rather than inflation-hit roubles.

On the other hand, their pay can be docked if anything goes wrong. In 1995 *Mir* commander Gennadi Strekalov was stripped of his benefits after he refused to take part in an EVA that he considered to be unsafe. Once he was back on Earth Strekalov contested the decision and succeeded in regaining some of his lost salary. A similar situation arose with Vasili Tsibliev and Alexander Lazutkin, the *Mir* crew held responsible when a Progress cargo ferry crashed into the station. They reportedly received 70 per cent of their fee on their return to Earth, but had to await the outcome of an official enquiry before their final salaries were determined. Not surprisingly, this system has been criticized for increasing safety risks by placing additional stress on the crew.

## SPACE TOURISM: GETTING AWAY FROM IT ALL?

Access to space will not always be the preserve of the professionals. As launch vehicles become safer and cheaper, anyone with the funds should be able to buy a ticket for the ride of a lifetime. However, apart from the risk of an untimely death during a launch explosion, there will be other drawbacks to overcome before space tourism really catches on. For example, floating in zero G may be fun, but throwing up in the cabin will be a definite problem for all concerned. Nor are there likely to be alcoholic beverages or after-dinner cigars on hand for those travellers who tire of wondering which way is up, who want a rest from hitting their heads on the ceiling or who become bored with watching the world through the window.

The views and experiences of Japanese journalist Toyohiro Akiyama may give a taste of what space travel will be like for ordinary people. Akiyama's 8-day adventure began with an ascent in a Soyuz rocket reminiscent of 'riding a dump-truck down a rocky road'. Things did not improve when he succumbed to space sickness. 'The commander told me to stop looking out of the window, and that helped,' commented the ailing cosmonaut.

His first words back on terra firma expressed a sense of relief that the whole episode was over. 'I'm hungry. I want to eat something delicious, have a beer and a cigarette. I've come back to Earth full of desires.' Then, too weak to stand, the reporter simply grinned and added: 'The air tastes good.'

# 2
# LIFT OFF!

Space missions may last for anything from a few hours to more than a year, but the success or failure depends largely on what happens during the few brief minutes of final countdown, launch and ascent. Seated on top of the biggest fireworks ever made, the crew can only watch, wait and pray that the army of designers, engineers and assembly workers knew what they were doing when they put together the complex piece of machinery on which their lives depend.

## THE LAUNCH SITES

Only two places in the world have experienced the drama and excitement associated with sending humans into space. Both Cape Canaveral, on the east coast of Florida, and Baikonur in Kazakhstan started out as military bases, constructed to send ballistic missiles against the enemy on the other side of the world. Top secret military missions are still conducted from these sites, but their main claim to fame now rests on their links with the civilian endeavour to conquer space.

Baikonur Cosmodrome lies in a semi-arid steppe region, noted for its scorching summers and bitterly cold winters. Its name is a legacy of Cold War secrecy, for the former Soviet Union deliberately named the launch site after a small mining town, about 180 miles southwest of the true location near Tyuratam. This deceit was soon uncovered by Western intelligence agencies, but the pretence continued and the name has been retained. Equally secret was the nearby town of Leninsk, built to house the numerous soldiers and engineering, design and support staff.

Construction began in 1955 when the military decided to move most of the rocket research away from Kapustin Yar, west of the Caspian Sea. The new site was close to Tyuratam railway station on the single track line from Moscow to Tashkent. A 17-mile spur was laid into the surrounding countryside. At the end of the line the engineers dug a deep pit and built Pad 1, the starting point for the first ballistic missile launch, the first satellite launch and the first manned launch. Baikonur has since grown to encompass a huge area stretching about 90 miles west–east and 60 miles north–south.

The site was off limits to Westerners until President de Gaulle broke the ice in 1966, but even during the joint Apollo–Soyuz Test Project, US astronauts were flown into the centre at night, installed in a hotel, allowed to visit only the launch pad and flown out after dark.

Since Baikonur is still controlled by the Russian Military Space Forces and will continued to be so until at least the year 2000 (when it should be handed over to the Russian Space Agency), gaining access to the site can be difficult. Visitors and cosmonauts usually fly into the airport at Leninsk (now confusingly renamed Baikonur Town). The cosmodrome itself has a runway and is also served by a dense network of railway lines and crumbling roads. Curiously, although the area is largely uninhabited, camel-owning nomads can still be seen wandering under the rockets' flight paths.

Rocket stages are flown from the factories to the cosmodrome, where they are joined and tested inside a giant vehicle-assembly building. For larger hardware, such as space station modules, a lengthy cross-country rail journey is required. A few days before the launch, the rocket is transported by rail to the pad, where it is tilted upright and filled with fuel. For a manned Soyuz flight, the cosmonauts arrive about two hours before launch and use an elevator on one of the service trusses to reach the capsule on the rocket's nose.

The break-up of the Soviet Union means that the cosmodrome now lies in the Republic of Kazakhstan, and much to their annoyance the Russians are obliged to pay rent for the use of facilities built largely with their money and expertise. Meanwhile, as the number of launches declines, so too do the facilities at Baikonur and Leninsk. Although commercial alliances with western companies are encouraging the development of modern hotel facilities, the town is still noted for its unannounced power cuts and water shortages. Buckets of water and candles are provided in hotel rooms in case of emergencies. The complex has also been plagued in recent years by a falling population and occasional riots by disaffected conscripts.

Cape Canaveral, on the other hand, has become a thriving tourist attraction. Its fame reached a peak in July 1969 when an estimated one million people poured into the area to witness the launch of *Apollo 11*. Even today, after more than 100 manned flights, thousands of people still flock to the nearby beaches and parking areas to witness the thrill of a launch. Bus tours from the visitors' centre cover the old missile pads used by the early astronauts, most now dismantled and rusting in the salty air, and the two shuttle pads, 39A and 39B.

While the Cape has all the buildings and facilities expected for a launch centre – control room, vehicle processing room, fuel depot and so on – the most awe-inspiring of them all is the vehicle-assembly building, originally constructed for the Saturn V rocket and now used for vertical mating of the shuttle, its fuel tank and two solid rocket boosters (SRB).

Two runways are available for shuttle landings and for use by astronauts flying in and out. Large pieces of hardware, such as SRBs, are brought from the factory by barge. Specially designed ships are also used to collect the SRBs after they run out of fuel and parachute into the sea.

Largely forgotten now is the second shuttle launch centre at Vandenberg Air Force Base on the coast of California. The complex was intended for military shuttle flights that would fly in near polar orbits, ideal for spy missions. Nearing completion when *Challenger* blew up in 1986, the pad was placed in mothballs and recently modified for the launch of small commercial rockets.

## PRE-LAUNCH RITUALS

Seven people have been killed during the ascent into orbit, and every space traveller must come to terms with the possibility that his or her life may come to a violent end while perched precariously on top of thousands of tons of highly explosive fuel. Preparing to meet one's maker is a personal process, and each astronaut copes with the possibilities in his own way. However, as with many stressful or dangerous occupations, a number of rituals have grown up to help keep the demons at bay.

One of the most unusual of these dates back to that April day when Yuri Gagarin boarded the bus for the short trip to the launch pad. Halfway there, the world's first

*Shuttle* Columbia *lifts off from pad 39B at Cape Canaveral on 19 November 1996.*

spaceman, future hero of the Soviet Union, suffered a case of acute nervousness. The bus screeched to a halt and the pressure-suited cosmonaut clambered down the steps to relieve his aching bladder. Ever since, crews have paid homage to the pioneer by disembarking on the way to the launch pad to lubricate the wheels of the bus. The few women who have flown on Russian rockets have been excused this particular ritual, with no apparent harmful effects.

Gagarin became such an inspiration to his colleagues that those who follow in his footsteps always visit the small wooden house near Pad 1 where he and Titov, his back-up, attempted to sleep the night before his flight.

Other pre-launch traditions based on the events of that historic day have become established in Russia. On leaving their Leninsk hotel room for the last time before a flight, the crews imitate Gagarin by signing their names on the door. Helen Sharman recounted how her crew were shown *White Sun in the Desert*, the film Gagarin watched on the night before his flight, and how she and her back-up, Tim Mace, were taken to the site where the hero died in a plane crash.

After donning their pressure suits, the cosmonauts sit in isolated splendour behind a glass screen, a quarantine procedure designed to prevent them from carrying an infection into orbit. They endure a series of questions and photo calls from reporters eager to know the details of their training and forthcoming mission.

The final ritual takes place before they board the buses. The crew assemble on the parade ground to declare their readiness to fly. Carrying their portable suit ventilators, they position their feet on three small white squares painted on the tarmac. The commander, standing in the centre, pronounces the famous words: 'My crew and I have been made ready and now we are reporting that we are ready to fly the mission.'

The chairman of the state commission replies: 'I give you permission to fly. I wish you a successful flight... and a soft landing.'

The prime crew then board a blue bus, distinguished by its licence plate 01 and a

*Members of the STS-65 crew practise climbing into a launch pad emergency slide-wire basket at pad 39A. From left to right: Rick Hieb, Chiaki Mukai, a Japanese crew member, and Don Thomas.*

lucky horseshoe; a second vehicle follows behind with the back-up crew. On arrival, the chosen team members clamber up some steps to the elevator and turn to wave farewell to the assembled crowd.

Once the mission has been completed, the crews return to Star City and place flowers at the foot of Gagarin's statue in the square that bears his name.

While the Americans have no equivalent to the hero worship accorded to Gagarin, they also have a pre-launch routine. Billeted in modest quarters at the Kennedy Space Center (KSC), the crew are separated from all but their immediate families and a few NASA officials and astronauts to limit possible germ contamination. On launch day wake-up time is about 5 hours before the scheduled liftoff time.

For many years, breakfast involved a 'high residue' banquet of filet mignon steak, eggs, orange juice and tea or coffee, which was eaten not only by the crew but also by their close friends and a few select members of the astronaut team. Today, crews eat more normal fare.

After breakfast they report for a final medical checkup and the attaching of body sensors and donning of pressure suits. All available pockets are filled with potentially useful items such as pens, pocket knives, sunglasses, calculators, flashlights and food-sticks. Personal belongings are collected and kept in a safe place ready for their return, together with visas ready to be sent to countries where the shuttle may make an emergency landing. All that remains is a stroll to the waiting bus – a modified mobile home that can seat up to 10 people – waves to the reporters, and a prayer that the launch will take place on schedule.

## FROM BALLISTIC MISSILE TO SHUTTLE

Humans would not have travelled in space had it not been for nuclear bombs and ballistic missiles, and even today, Russian cosmonauts are carried into orbit perched on top of a rocket that was originally designed to despatch nuclear warheads to the United States.

As might be expected for complex vehicles assembled from thousands of parts, each built by different contractors under a system that favoured the lowest tender, 100 per cent reliability is not a realistic expectation.

In the absence of a perfect flying machine, various methods to keep the crew alive have been devised. The Vostok cosmonauts were strapped into an ejection seat that could be used after re-entry and as a launch escape system. However, the system was removed on the Voskhod follow-ups to make room for a three-man crew or an inflatable airlock. Fortunately, the cosmonauts' luck held. If anything had gone wrong, there would have been no escape.

Ejection seats equipped with a rocket catapult were also built into the two-man Gemini craft. This dangerous contraption was designed to throw an astronaut 1000 feet from an exploding Titan booster in just 6 seconds. A small ballute, a hybrid balloon-parachute, acted as a stabilizer.

More sophisticated versions were later installed in the shuttle during its initial four flights with two-man crews. Since they were not feasible when larger crews began to fly, they were disarmed for STS-5 and not fitted on all subsequent flights. Ejection seats were also developed for the Russian *Buran*, but never used.

Shuttle crews now rely on a rapid evacuation of the cabin if danger threatens during countdown. If a fire occurs on the pad while the vehicle is still grounded, the crew are expected to exit through a side hatch, jump into wire baskets and slide down overhead cables to a concrete bunker.

Probably the nearest they came to being used in a crisis was during a launch abort on the twelfth shuttle mission. Two of the three main engines started but were shut down by the on-board computer when a fuel valve failed to open properly. Residual fuel ignited fires on the pad, and these took some time to extinguish. For a while it looked as if the slide-wires would be used.

Once the vehicle leaves the pad, the crew largely rely on the skills of the commander and pilot. During training they practise a variety of launch abort procedures, including the separation of the shuttle from its orange fuel tank and an emergency landing on the nearest, most suitable runway or ditching in the sea. After the *Challenger* disaster an additional emergency evacuation procedure was introduced, by which the crew slid out along a pole extended from the side hatch and parachuted into the ocean. It is, however, generally accepted that there is no way a crew will survive if a serious malfunction occurs while the fiery SRBs are still attached.

Almost all the expendable rockets built to carry humans into orbit have been equipped with an escape system. This usually takes the form of several solid fuel escape motors attached to the capsule containing the crew. These can be fired automatically or on a command sent manually by mission control if a serious malfunction occurs on the launch pad or during the first stages of lift-off.

Once the engines fire, they cannot be shut off. The escape stage pulls the crew cabin free of the faulty booster until it is safe for parachutes to be deployed. The parachute will, it is hoped, have time to slow down the capsule before it strikes the ground with a sickening thud.

Not surprisingly, cosmonauts are offered no training to prepare themselves for such an eventuality – simply given a verbal run-down of the situations that might trigger such a drastic action, and a description of the noise, bone-shaking vibrations, crushing acceleration, probable unconsciousness and hospitalization that will follow.

## ABORT!

Only one crew has ever experienced the misery of such a launch-pad abort. On 27 September 1983 Vladimir Titov and Gennadi Strekalov were expecting to begin a routine flight to the *Salyut 7* space station. With just 90 seconds to go, fire broke out at the base of the Soyuz booster when a propellant line failed to close. The flames spread, enveloping the entire lower section of the rocket. Helpless in their cabin far above the ground, the crew could only lie back and wait either for the rocket to explode under them or for the abort system to be activated and blast them free. They would not have felt more confident had they been aware that the flames had burned through the wires of the automatic abort system, making it inoperative.

Twelve long seconds passed before, in the nick of time, two launch controllers managed to abort the mission by sending radio commands from their blockhouse. The Soyuz T-10 escape engines fired, pulling the spacecraft free. Beneath the rapidly accelerating ship, the booster toppled like a felled tree and disappeared in a huge ball of fire. The crew experienced the fairground ride to end all fairground rides as they were pressed back in their couches by 17G, and $5\frac{1}{2}$ minutes later their descent cabin made a bumpy touchdown in a field. Both escaped unscathed. Some $2\frac{1}{2}$ miles away, the launch pad they had so proudly occupied such a short time ago was a smouldering ruin.

## LAUNCHING A SHUTTLE

Launching a complex space vehicle such as a shuttle is a precise operation with several major steps. The first step in the final launch sequence is the ignition of the shuttle's three main engines at T− 3.46 seconds. By T = 0 these engines have reached 90 per cent thrust. If they are burning properly and reach full thrust within 6 seconds, the two SRBs ignite at T+ 3 seconds, followed at once by the release of the bolts that shackle the orbiter to the pad. This is the point of no return.

Within 4 seconds of leaving the pad, the shuttle begins to pitch over, a manoeuvre that carries it out over the Atlantic Ocean. The space shuttle main engines (SSME) are temporarily throttled to reduce stresses on the orbiter. Just over 2 minutes into the mission, the SRBs run out of fuel and are jettisoned. They parachute into the ocean and are towed back to shore, where they can be refurbished and reused. The shuttle continues to climb eastward over the ocean, burning its supply of liquid hydrogen and oxygen. Some $8\frac{1}{2}$ minutes after lift-off, the huge external fuel tank, now empty, is released and allowed to burn up during re-entry. Final insertion into orbit is achieved by firing the smaller engines of the shuttle's orbital manoeuvring system (OMS) once or twice.

As always when dealing with such complex machinery, mission controllers and crew have to be prepared for something to go wrong. A number of abort modes are also built into the launch and ascent profile:

❶ A major problem early in the flight, such as shut-down of one or more engines, requires the pilot to 'return-to-launch-site' by pitching around and thrusting back towards the Cape until within gliding distance of the runway.

❷ Loss of one or more main engines midway through the ascent would force a 'trans-Atlantic abort landing' at one of the runways in Gambia, Morocco or Spain.

❸ Early main engine shutdown may allow an 'abort-once-around' – i.e., one orbit of Earth with a return to Cape Canaveral.

❹ Partial loss of main engine thrust late in the ascent will allow the shuttle to 'abort-to-orbit' – i.e., reach a minimal 120-mile orbit with the aid of the OMS engines.

## Earth to Space in 8½ Minutes

Andy Thomas described the STS-89 shuttle launch as seen from the mid-deck of shuttle *Endeavour* on 22 January 1998:

I was one of the first to strap in on the mid-deck, so I had a long wait on my back before the engines were lit. The big difference between sitting on the mid-deck, versus the flight deck, as I had done for my first flight, is that I had no window to look out of and watch the world recede. But not having an outside view lets your imagination provide the imagery, and this can give you an emotional rush, possibly even more than seeing.

The weather had been questionable that day ... . But a few minutes before launch, all the launch controllers were polled by the launch director and each gave a 'go' for launch. The Control Centre then called us to start the auxiliary power units that provide the steering hydraulics and we could hear the units spinning up to speed deep below us... then came the call to close and lock our visors, and to initiate our oxygen flow, a protection in the event of a depressurization during the climb-out... The three of us on the mid-deck shook hands together and wished ourselves well for the flight. Then the cabin became quiet. ...

At 6 seconds before launch a deep rumble started shuddering the orbiter as its three engines were ignited and run up to full speed. ... But we were still firmly bolted to the ground with eight very large explosive bolts so the engine thrust made us lurch over, giving us the eerie sense of falling forward. Suddenly, with the 6 seconds counted away, there was a thundering roar with massive vibration and shaking as the solid rockets were ignited, the hold-down bolts exploded, and we were driven off the launch pad and upwards into the sky. You did not need a window to know what was happening.

... We did a roll manoeuvre to adopt the correct flight orientation, causing us to feel the whole cabin spin around while being shaken by the 8 million pounds of thrust accelerating us forwards. The flight deck crew called out the rapidly changing speeds and altitudes every few thousand feet. We could feel the engines throttle back to prevent over-stressing the vehicle in the denser part of the atmosphere, then they came up to full speed again driving us once more back into our seats. But burning fuels at a combined rate of 12 tons a second quickly depletes even the huge solid boosters and after about 2 minutes we could feel a noticeable drop in acceleration. It actually felt as if we were slowing down and pitching forward, but ... we were still climbing upwards very swiftly.

There was a jolt as the solid boosters, having now become just dead weight, were explosively separated at about 70,000 feet above the Earth. Then the engines, feeling the sudden lost of mass, pushed us harder forward. Now, the changing altitudes were called out not in feet, but in miles, and the changes in speed in thousands of feet per second as we flew through the upper atmosphere at hypersonic speeds.

At an altitude of 50 nautical miles, the traditional beginning to space, we congratulated the novice crew members on reaching space, and I shook the hand of the cosmonaut Salizhan Sharipov sitting to my left. The accelerations steadily increased as our speed became faster and faster, and at 3G acceleration the engines were again slowed down to prevent excessive loads. But even that meant that we would add nearly 4000 miles per hour to our speed every minute. And at that acceleration, our speed soon reached 17,500 miles per hour, orbital velocity. We were at orbital altitude, and the engines shut down. The sudden loss of acceleration again gave us the false sensation of pitching forward. But in reality we were coasting in space and were now weightless.

The only one of these ever to be put into practice was an abort-to-orbit on STS-51F, when one SSME was shut down after sensors showed it was overheating. The crew were obliged to head for a low orbit using *Challenger*'s two remaining engines. Severe shock set in when a second motor appeared to be overheating. An emergency landing in Spain or, at worst, a ditching in the Atlantic, seemed likely. Fortunately, a ground engineer

*The shuttle* Atlantis *clears the tower at the start of mission STS-71, the 100th US human spaceflight.*

rightly suspected that the fault lay with the sensor and instructed the crew to countermand the impending shut-down.

Use of the shuttle's manoeuvring system engines and aft thrusters enabled the orbiter to struggle to a height of 170 miles, 70 miles short of its target altitude. Some of the Spacelab experiments were adversely affected, but the week-long mission was saved.

## THE *CHALLENGER* DISASTER

The perception that the shuttle is simply a space transportation system, an orbital version of a jet airliner, is, and always has been, flawed. NASA fed the popular illusion of a cheap, reusable and, above all, reliable space vehicle by declaring it operational after just four 'test' flights. The danger of this type of thinking was brought home in force by the tragic loss of all seven crew in the 1986 *Challenger* disaster.

The programme had been running into problems for some time. The launch of the previous mission, STS-61C, had been scrubbed four times because of adverse weather or technical problems. Mission 51L was in danger of repeating the performance. Severe cross-winds had already postponed the launch on 27 January, then came a night of unprecedented sub-zero Florida temperatures. As the Sun rose on 28 January, its rays glinted on rows of icicles smothering the gantry.

When they learned of the extreme conditions, engineers from SRB manufacturer Morton Thiokol expressed concern about a failure in the rubber seals, known as O-rings, which could release red hot gases between segments of rocket casing. Under pressure to maintain the schedule, mission managers overruled warnings that the shuttle's solid fuel motors had never flown with the mercury so low. Despite the ice and record cold, the order was given for the countdown to commence.

As the Sun rose, the air temperature climbed to 2°C, and *Challenger* seemed set to go. As the main engines ignited, a cheer went up from the watching crowds. Seconds later the SRBs lit up in a searing white sheet of flame. Unseen by anyone, a telltale puff of smoke blew out from a seal on the right-hand SRB. The crew's fate was already sealed.

Little more than half a minute had passed by the time *Challenger* began to be buffeted by high altitude wind shear. The orbiter's guidance system compensated as the winds continued over the next 27 seconds, then the engines returned to maximum thrust. By this time, a tiny flame had appeared towards the lower end of the right-hand SRB and started to spread. Hot gases were leaking from the damaged joint.

It was just a matter of time. Deflected by the air flow, the flame turned like a blowtorch against the main fuel tank and the strut that attached the tank to the flawed SRB. At 64.7 seconds into the flight, the fire ate its way through the tank's skin, allowing hydrogen to escape. Less than 8 seconds later, as the shuttle struggled to correct the forces that were throwing it off course, the lower strut gave way. Immediately, the hydrogen tank began to collapse, while the loosened SRB swung round and pierced the oxygen tank.

Within milliseconds it was all over. The gases streaming from the punctured fuel tank ignited. At a height of 46,000 feet above the ocean and travelling at around 1400 mph, *Challenger* was surrounded by a white plume that expanded to envelop the vehicle. The shuttle broke into several pieces as its reaction control system ruptured and caught fire. For a few seconds, launch commentator Steve Nesbitt continued to read off the altitude and velocity, until he realized what everyone was staring at. 'Flight controllers looking very carefully at the situation. Obviously a major malfunction,' was all he could say.

Throughout the brief, dramatic events, there was no evidence of alarm from the crew. Only as their vehicle broke apart around them did they try to respond to the sudden danger. Commander Scobee just had time to open his radio channel but was cut off before he could speak. Pilot Michael Smith, suddenly aware that something was terribly wrong, exclaimed, 'Uh, oh.' Some of the crew activated emergency oxygen masks, but with no effect. The craft's shattered remnants plummeted 9 miles into the sea, scattering over hundreds of square miles.

A Presidential commission was set up to investigate the worst accident in the history of human spaceflight. Once the official enquiry

*Icicles hang from the launch gantry on the morning of 28 January 1986. During the previous night the temperature plummeted to well below freezing, but despite the record cold temperatures, launch managers decided to give a 'go' for the launch of* Challenger.

discovered the cause of the *Challenger* accident – hot gases inside the SRB bypassed the internal insulation and O-rings to melt through a steel joint – most of the effort to restore confidence and improve safety was concentrated on redesigning the SRB. A third O-ring and a J-seal were added to ensure that no leakage can occur when pressure builds up inside the motor. To combat cold weather, heaters were placed on the outside of the booster so that the O-rings would remain warm and flexible.

The SSMEs were upgraded, including work on the high-speed turbo-pumps and the main engine controllers. Rescue procedures in the sea, along the Florida coast and on land were also reviewed.

Some attempt was made to improve the chances of the crew surviving a similar launch problem. Coverall-type flight suits for astronauts were out, orange partial pressure suits were back in. A bail-out pole was added to the mid-deck so astronauts could be guided while jumping/bailing out without contacting the wing or engine area. Whether they would be able to avoid the inferno belching out of the SRBs was another matter.

## DOCKING

In the early days each spacecraft functioned as a separate entity, but it was soon recognized that launching a spacecraft into orbit was not enough. The future of manned spaceflight depended on being able to transfer cargo and crew from one ship to another. Some way had to be found of linking up two craft as they travelled around Earth at 17,500 mph.

The ability to locate and track another space vehicle and then manoeuvre into position for a docking had to wait until the development of small computers and radar. The crew of *Gemini 8*, Neil Armstrong and David Scott, were the first to achieve an orbital union when, on 16 March 1966, they pulled in at the end of an Agena rocket. But triumph almost turned into disaster.

The combined craft began to roll. Convinced that the problem lay with the Agena, Armstrong broke free, only to discover to his horror that the Gemini's rate of spin began to spiral out of control. As the rotation grew to one revolution per second, the crew were in danger of blacking out. Armstrong's one remaining card was to shut down the main attitude control system and use the re-entry control thrusters to regain control. The ploy worked. The crew made an emergency splashdown less than 11 hours after their launch. No one cared about the curtailed mission. They were just glad the men had survived.

From that inauspicious start, the Americans went on to perfect the manual docking procedures that enabled them to win the race to the Moon. Although other orbital link-ups followed during the *Skylab* and Apollo–Soyuz programmes, they became rarities with the introduction of the shuttle. Not until an agreement was signed with the Russians to share the *Mir* space station did the shuttle dock with another vehicle.

Curiously, despite their more advanced electronics, US spacecraft have always used pilots to carry out orbital unions. It has been left to the Russians to develop automatic rendezvous and docking. The first of these manoeuvres took place between *Cosmos 186* and *Cosmos 188* on 31 October 1967. This was followed on 16 January 1969 by the first docking between two manned craft, *Soyuz 4* and *5*, an event that preceded the first link-up between two Apollo craft by about 7 weeks.

Over the past three decades the Russians have built up a tremendous amount of experience of automatic dockings through their unmanned Progress and manned Soyuz spacecraft. Of 42 first-generation Progress launches, all successfully docked with space stations using the Igla automatic approach and docking system.

The first serious problems came with the upgraded Progress-M, which used a more advanced Kurs system. One of the most critical of these occurred on 23 March 1991. Following an unexplained shut-down of the approach system two days earlier, when Progress M7 was just 1500 feet from *Mir*, a second attempt was made to bring the two craft together.

This time all seemed well as the Progress moved in towards the docking port of the *Kvant 1* module. However, as the gap closed, ground controllers realized that it was not aligned. Emergency commands to abort the approach were beamed to the Progress in the nick of time. As the *Mir* crew watched in stunned amazement, the 7-ton ship turned aside and passed within 40 feet of the station, miraculously gliding past all the aerials and solar panels. As one TV reporter commented: 'Thank goodness it approached the *Mir* station on the side without the two 20-metre [66-foot] modules.' It was later revealed that the problem lay with a faulty Kurs antenna on the *Kvant 1* module, which had been damaged when a cosmonaut accidentally kicked it during a spacewalk.

Ironically, most of the docking problems have involved the manned Soyuz craft. The first attempt to dock two Soyuz in 1967 was cancelled when pilot Komarov ran into trouble and died during re-entry: the second craft with which he was due to rendezvous never got off the ground.

The first attempt to dock with a space station was only partially successful. *Soyuz 10* linked up with *Salyut 1*, but the ships were not pulled together in an airtight seal and so the crew never saw the inside of the station. Other failures followed at regular intervals.

The most dangerous of these involved the crew of Soyuz T-8, Gennadi Strekalov, Alexander Serebrov and Vladimir Titov. Once in orbit, the men soon realized that their mission was in jeopardy when the antenna of the Igla rendezvous system got stuck in the wrong position during deployment. Attempts to shake it free failed. Mission rules decreed that the flight should be cancelled and the crew should head for home at the earliest opportunity.

On this occasion the rules were ignored and the crew were given the go-ahead for an unprecedented 'eyeballs-only' attempt, using a special drift indicator. With the aid of instructions from the ground, the crew slowly approached their target, all the time trying to estimate their distance from *Salyut 7* by studying its apparent size on a screen divided into grid squares.

As they tentatively drew nearer, they prepared for darkness by switching on the searchlight. By the time their ship entered orbital night, the crew had drifted out of range of mission control. Left to their own devices, they refused to wait for daylight. Titov described what happened next:

*We hurtled on and the distance to the station was 280 metres [925 feet]. We controlled the station's position on the screen. We could feel that the rendezvousing speed was high. I fired the deboosting engine. The distance was 160 metres [530 feet]. Still the speed remained quite high. At night, when the distance and speed of rendezvousing are difficult to assess visually, the danger of collision is quite real. I fired the engine to make the spacecraft descend a little. We flew by the station.*

Some years later Strekalov revealed that this matter-of-fact description cloaked a near-disaster and called the experience 'a brush with death', with the Soyuz sweeping past the station at high speed, barely avoiding its projecting solar panels.

# 3
# LIFE AFLOAT

Humans take gravity for granted. We accept that Nature has equipped us with a skeleton for strength and muscles for movement. Without these, we would collapse into a gelatinous mass. Had we evolved in outer space, our shape and structure would be entirely different. Bones, muscles and physical strength are almost redundant in a weightless environment.

Motion is a case in point. Any astronaut can turn somersaults that would make the most adept circus performer green with envy. The merest touch against a solid surface can send an astronaut careering in the opposite direction across a spacecraft cabin. This movement will continue until another force is applied or the astronaut comes into contact with a solid object. Rookie astronauts are easily identifiable by the numerous cuts and bruises inflicted by their uncontrolled cavortings.

*Every which way is up. European Space Agency astronaut Thomas Reiter looks 'down' on his smiling Russian colleagues, Yuri Gidzenko (right) and Sergei Avdeyev inside the* Mir *docking module.*

## SPACE SICKNESS

Almost without exception, space travellers consider life in zero gravity to be a pleasant experience, once the first few days have been negotiated. During the early adaptation period, many astronauts suffer from space sickness, a feeling of nausea similar to travel sickness. Other symptoms include dizziness, headache, apathy, inability to eat and cold sweats.

The first to suffer from this unpleasant condition was Gherman Titov, the second man to orbit the planet. During his 24-hour flight, Titov came close to vomiting as he floated in the relatively spacious Vostok cabin. Three years later, *Voskhod 1* cosmonauts Boris Yegorov and Konstantin Feoktistov reported feeling upside down throughout most of their short mission. Head movements led to brief dizzy spells, while Yegorov also exhibited a loss of appetite and 'unpleasant feelings in the stomach', which reached a peak 7 hours into the flight.

While some of their Soviet counterparts were suffering from space sickness, the early American pioneers sailed effortlessly around the globe. Doctors were puzzled by their apparent immunity. Only later, as American spacecraft became larger and astronauts floated freely around the cabin, did the same symptoms manifest themselves. It became apparent that space sickness was related to conflicting signals reaching the brain from the vestibular system in the inner ear. Deprived of gravity and the normal visual up and down reference points, the brain becomes confused, though why this should result in nausea remains a mystery.

Despite attempts to prepare astronauts for weightlessness

through training flights on the 'Vomit Comet', space sickness remains a hazard that can afflict a crew during the first few days of a mission, and it has been known to persist for a week. Such disability can have serious implications for a crowded work schedule, and on at least one occasion, an astronaut has felt too ill to participate in an EVA. Aware of this problem, mission planners tend to avoid scheduling spacewalks early in a mission.

One way astronauts can minimize the problem is to use drugs such as scopolamine dexedrine or Phenergan, but this is usually a last resort, since they tend to induce drowsiness. Certainly, vomiting is something to be avoided in zero gravity, since the stomach contents could soon spread around the cabin. The space toilet is not designed to handle such situations, so rookie astronauts tend to carry sick bags with them at all times.

Space sickness offers a psychological challenge to astronauts who have developed a macho persona through their careers as military men and test pilots. When William Pogue threw up, the crew conspired to dispose of the evidence down the waste airlock and to pretend nothing had happened. They had forgotten that their conversations were being taped and replayed back to mission control. The result was the first public reprimand issued to a crew during a mission.

A reluctance to discuss the intimate secrets of one's digestive system has continued to the present. A typical comment came from veteran Paul Weitz after the sickness problems of the STS-5 crew were given wide publicity: 'I think it's between me and my doctor and is nobody else's business.'

## ADAPTING TO LIFE OFF EARTH

Space sickness apart, astronauts find zero gravity a liberating experience. Many astronauts have commented on the increase in comfort associated with weightlessness. Although confined to a couch for almost 5 hours, the first American in orbit, John Glenn, experienced much less discomfort than he had anticipated.

*Being suspended in a state of zero G is much more comfortable than lying down under the pressure of 1 G on the ground, for you are not subject to any pressure points. You feel absolutely free.*

Despite the brevity of his flight, Glenn found he soon adapted to life off Earth. 'Without even thinking about it, I simply left the camera in mid-air, and it stayed there as if I had laid it on a table until I was ready to pick it up again.'

*Mir* space station resident David Wolf had a different viewpoint about getting used to his weightless environment. 'Every item you touch just floats off if you don't Velcro it, or strap it down, or bungee it in place … . I keep finding myself with too many things in my hands and no way to put them down.'

This wonderful equilibrium can be upset if the engines or thrusters have to be fired. 'We just weren't prepared,' recalled *Apollo 14* crewman Stu Roosa. 'Things were flying back by our heads that we just hadn't fastened down.'

## THE SPACE GYM

Early in the space programme it was discovered that a number of changes take place in an astronaut's body. In particular, bone and muscle strength deteriorate in the absence of gravity, and this increasing physical weakness may become life-threatening when the body undergoes deceleration forces during re-entry and returns to normal gravity.

Other physiological changes involve the circulatory system. In normal conditions, gravity pulls body fluids towards the legs, forcing the heart to pump harder to carry

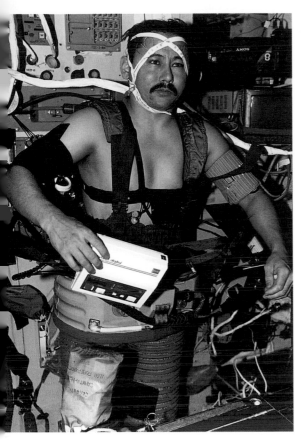

*Kazakh cosmonaut Talgat Musabayev wears the Chibis trousers during preparations to return to Earth after many months in space aboard the* Mir *station.*

blood and oxygen to the brain. In weightlessness, these fluids travel freely around the body. Released from the burden of gravity, the heart muscles become lazy and weak. When the astronaut returns to Earth, the heart is no longer in a condition to resume normal service.

Among the side-effects of this shift in body fluids is an increase in puffiness around the eyes. 'One feels this strange fullness in the head and this sensation of having a cold and the nasal voice, and one sees the puffy look on the faces of his fellow crewmen,' commented *Skylab*'s Joe Kerwin. 'One can almost see the fluid draining out of the legs; one looks at his partners, and their legs are getting little and skinny like crow's legs.'

In addition, astronauts become one or two inches taller as their spines stretch. Such an increase in height has to be allowed for when considering the design of a snug-fitting pressure suit. Andrew Thomas got off to a bad start on his long duration mission to *Mir* when he complained that his Russian Sokol pressure suit did not fit. Deputy flight director Viktor Blagov sniffily complained that the newcomer was being 'capricious'. 'One ought to meet this in a Russian way and, as all our cosmonauts do, just put up with it for a few minutes.' The problem was solved by adjusting the suit's built in straps.

The best antidote is to employ the muscles by providing heavy work loads. In the cramped capsules of the early manned programme, space was at a premium, so ingenious ways were found to offer forms of exercise to the couch-bound occupants. During Glenn's 1962 orbital flight, he was asked to pull on a bungee cord attached under the instrument panel. He later wrote:

*I gave it one full pull per second for 30 seconds to see what effect exercise would have on my system under a condition of weightlessness. ... It made me tired. My pulse went up from 80 beats per minute to 124 beats in 30 seconds, but it returned to 84 beats per minute within a couple of minutes.*

Today, a strict exercise regime is built into the daily flight routine. On a typical shuttle mission, astronauts are expected to exercise for at least 15 minutes each day. This is increased to 30 minutes on longer missions, while on a space station 2 hours a day is the norm.

One of the most used exercise machines is a treadmill. The shuttle version is a Teflon-coated aluminium sheet on a roller with a frame that locks into holes in the floor. Astronauts hold themselves in place with a strap attached to its base. Other machines installed on the shuttle and space stations include a rowing machine and an exercise bicycle known as an ergometer. The pedals on the ergometer can be used for exercising both the arms and the legs. On the latest model, the astronaut is expected to pedal while lying horizontal to the floor rather than in an upright position.

Exercise can have its downside, as Norman Thagard found when the chest expanders he was using snapped and the spring caught him above an eye. A less strenuous way to strengthen the heart and muscles is to wear an elasticated 'Penguin' suit.

As the rigours of re-entry approach, astronauts try to simulate gravity by decreasing pressure on their lower body. On the *Skylab* missions of the 1970s this was done by climbing into a cylinder reaching up to the waist, but today the Russians use a special pair of Chibis trousers. In each instance, some of the air is sucked out, creating a partial vacuum. This causes blood to pool in the legs, increasing the heart's work rate. The method has to be used with care, however. Loss of blood volume as a result of prolonged spaceflight, combined with a reduction in blood reaching the brain, caused the *Skylab* astronauts to feel faint.

## KEEPING CLEAN

Cleanliness was almost nonexistent on early space missions when spare room was at a premium. Even today, shuttle astronauts have to make do with a wipe over using a moist, sanitized cloth. Fortunately, shuttle missions rarely last more than two weeks, so crews put up with the inconvenience. On longer trips, a shower would normally be considered as essential. For more than 20 years, space station designers have tried to meet this need, but with only modest success.

All Soviet space stations since *Salyut 3* have carried a shower, as did the American *Skylab*. However, in all cases they have proved to be unreliable and unpopular. On *Skylab* the astronauts could rarely be bothered to spend hours setting up the collapsible shower stall and stowing it away once more. The men also complained that their fireproof towels were rough and non-absorbent. 'Like drying off with padded steel wool,' growled Ed Gibson. They resorted instead to drying off with the station vacuum cleaner.

Cosmonaut Valeri Ryumin was not much more enamoured with the Russian model. He and colleague Vladimir Lyakhov usually attempted a shower once a month, but the thought of all the work involved was sometimes just too much. He wrote in his diary:

*You have to heat the water, in batches, no less. You have to get the shower chamber, set up the water collectors, attach the vacuum cleaner ... it takes nearly the entire day just for that shower.*

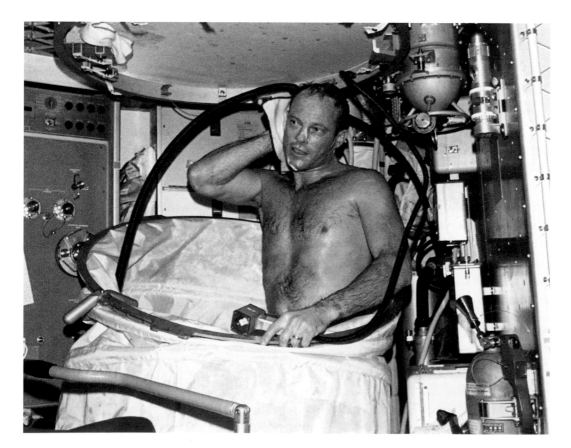

Skylab 3 *pilot Jack Lousma dries himself after a hot shower. The* Skylab *shower curtain was pulled up from the floor and attached to the ceiling. The water came through a push-button shower head which was on the end of a hose.*

A similar collapsible shower was part of the *Mir* core module launched in February 1986. A permanent cubicle was later added as a fixture in the *Kvant 2* module, but by early 1995 the crew were using it only as a sauna.

As on Earth, the method of shaving is a matter of personal preference. Indeed, some astronauts preferred not to shave at all, though they often made themselves more presentable before the recovery helicopter deposited them in the media spotlight. Although electric razors are easier to use than traditional safety razors, the fine hairs they produce are unpleasant to inhale, so shaving is recommended near to an air filter. A wind-up version is used on the shuttle, equipped with a vacuum device to capture the whiskers. Hair cuts are necessary only on longer flights and are best carried out with a vacuum cleaner at hand.

## THE AIR THEY BREATHE

Deciding on the most suitable composition and pressure of cabin air has always been a problem. On the first spacecraft, the Americans and Soviets plumped for entirely different solutions. Soyuz cabin atmosphere consisted of the usual oxygen-nitrogen mixture at sea level pressure of 14.7 psi (pounds per square inch). However, in Apollo the air was pure oxygen at a pressure only one-third that in Soyuz. These considerable differences caused a few headaches in the lead up to the Apollo–Soyuz mission of 1975. Cosmonauts wanting to visit Apollo would receive a painful and possibly fatal attack of the 'bends' caused by nitrogen bubbles in their body tissues.

Agreement was finally reached on the need for an airlock in which the atmosphere could be adjusted during transfer from one craft to the other. Once the atmospheres in the airlock and the destination ship were matched, the crew could open the hatch. Equipment was also built into Soyuz so that air pressure could be reduced during the period of joint flight.

The complex arrangement worked perfectly. Once in orbit, the oxygen content in the Soyuz was increased at the same time as the air pressure was lowered from 14.7 psi to 10 psi. After the two craft linked up on the third day, there followed the historic first handshake in orbit between space travellers from either side of the Iron Curtain.

## FIRE IN THE COCKPIT!

Pure oxygen has many advantages, but it has one major drawback – it is highly flammable. An outbreak of fire will become an inferno in a matter of seconds.

This is exactly what happened during a ground test of the *Apollo 1* capsule on 27 January 1967. Already disenchanted with the bug-laden spacecraft, veteran astronauts Gus Grissom and Ed White, along with rookie Roger Chaffee, slotted into their seats for a lengthy shake-down. The hatch was sealed and locked behind them. No danger was anticipated, since the rocket was empty of fuel. However, the command module cabin contained pure oxygen at 16 psi, above normal air pressure to prevent nitrogen from seeping into its atmosphere.

Frustration followed frustration as communications were frequently garbled. At one stage the session was put on hold when Grissom noticed a strange odour in the suit oxygen supply, but nothing amiss was found.

After 5½ hours in the couches, the countdown simulation seemed to be reaching its climax. At 6.20 p.m. a hold was called prior to the switch-over to fuel cell power. It was now dark outside and the crew were looking forward to a respite from their trials and tribulations. Then, at 6.31, a sudden power surge was recorded somewhere in the 20 miles of wiring surrounding the astronauts. Less than 10 seconds later, Chaffee reported, 'Fire, I smell fire.'

Immediately, the men began to initiate the emergency escape procedure, but it was too late. As Ed White tried to find the hatch handle and undo the six securing bolts, he and his colleagues were engulfed in a fireball. So dense was the smoke that the men became invisible on the TV screens. Then came the final, heart-rending cry, 'We've got a bad fire... we're burning up here!' followed by sounds of frantic movement, shouting and pounding on the cabin wall.

Just 16 seconds after the initial report, the capsule split apart in a violent explosion. Hampered by the searing heat and dense black smoke, the pad team struggled to reach the trapped men. Minutes passed before they were able to wrench the hatch free. A terrible sight greeted them. The astronauts' bodies lay among the blackened ruins of what had once been a spacecraft. They were welded to the cabin by a solidified glue composed of melted spacesuit, nylon netting, Velcro and insulation. Not until 7 hours after the incident could they be released. The official cause of death was asphyxiation caused by inhalation of smoke.

The multi-billion dollar Moon programme ground to a halt while an official enquiry sought an explanation for the tragedy. Three months later a 3000-page report scathingly attacked the sloppy procedures that had produced a death trap rather than a craft capable of flying to the Moon. An electrical spark was blamed as having been the immediate cause, but the report went on to catalogue 'numerous examples of poor installation, design and workmanship'.

The inevitable outcome was a complete redesign of the command module. Both the craft and the spacesuits were made as fireproof as possible. The pure oxygen atmosphere would be used only in orbit, and then at one-third of normal atmospheric pressure. During ground tests and launch countdowns, the cabin would contain a safer nitrogen-oxygen mixture while the crew breathed pure oxygen in their suits. Most significant of all was the redesign of the hatch. From now on, astronauts would be able to open the door and make their escape within 10 seconds instead of the 90 seconds required previously.

## FIGHTING FIRE IN ORBIT

A sudden outbreak of fire is potentially even more serious for astronauts trapped hundreds of miles above Earth than for the average householder down on the ground. A particular problem is that fires behave differently in zero gravity. Terrestrial fires exhibit tongues of flame as the hot, expanding gases rise. Soot particles are also swept along by the up-draught, creating the familiar cloud of smoke. In orbit, however, the absence of gravity means that convection currents cannot flow. Unless fresh oxygen is supplied by moving air currents, a fire will tend to burn slowly and become spherical in shape.

Not surprisingly, considering their limited extent and slow development, space fires are often difficult to detect. Just as worrying is the fact that they are surprisingly common and quite difficult to extinguish.

In November 1994 NASA admitted that five 'fire incidents' had occurred aboard the shuttle. All of them were caused by short circuits or electrical components overheating. Worse still, the agency admitted that, on each occasion, the smoke detectors failed to work and the crew became aware of the danger only after they smelled or saw smoke.

Fortunately, they were spotted at an early stage, enabling the crew to act immediately by switching off the circuits without having to resort to fire extinguishers.

If an orbital fire does take hold, how do you put it out? The obvious answer is to squirt water on the flames. However, appliances using compressed gas and water will not work in weightlessness. The gas and liquid simply mix together like gaseous mineral water. In addition, water droplets tend to float in zero gravity and have a nasty habit of seeping into electrical systems, causing short circuits. NASA has relied instead on aircraft-style extinguishers, which use halon gases. Six of these are installed on the shuttle. Although they produce acidic fumes as by-products, they are considered fairly safe at the low concentrations required to snuff out a fire.

Officials of both space superpowers have been reluctant to announce such outbreaks in public. In 1994 RSC Energia filed a hazard report denying that there have ever been any incidents involving fire aboard Russian manned space vehicles, even though published comments by cosmonauts had named about half a dozen such cases. When another fire occurred on *Mir* in October 1994, it was not immediately disclosed to the public by either Russian or American space officials. On this occasion, a lithium perchlorate cartridge caught fire, but it was quickly put out by Valeri Polyakov.

History was to repeat itself. The fire that occurred on 23 February 1997 drew its own blaze of publicity since it took place during a 'party' in *Mir*'s core module involving German Reinhold Ewald and US astronaut Jerry Linenger. The presence of these two visitors and their four Russian companions meant that *Mir* was unusually crowded at the time. Although there were two Soyuz craft docked at either end of *Mir* and available for an emergency return to Earth should the fire spread out of control, one of them was cut off by a wall of smoke and flames. For at least three of the crew, the only way to survive was to extinguish the fire.

The fire began in the *Kvant 1* module when the crew were using lithium perchlorate canisters to produce oxygen. Each candle is designed to burn as a back-up system when the main Elektron system is shut down or unable to cope. They give off the life-giving gas during combustion. On this occasion, the fourth one used that day ruptured, turning rapidly into a blow torch. The canister began spraying molten metal as the flames threatened to spread to the panels on the starboard side of the *Kvant* module. Several electricity cables also caught fire. Here is Linenger's description of the moment all astronauts dread:

*The fire took place just behind me in the* Kvant *module ... . Once the fire broke out, the master alarm went off, [and] smoke filled the station ... . Without getting that fire out there was no way to get to one of the Soyuz capsules.*

*... I did not expect smoke to spread so quickly. It was a magnitude about 10 times faster than I would expect a fire to spread on a space station. The smoke was immediate, it was dense. Where I was sitting I could see a shadowy figure of the person in front of me ... but I really could not make him out. Where he was standing, he could not see his hands in front of his face. ...*

*We immediately went to the oxygen-breathing device and without that I don't think you would have been able to breathe. When I first activated the first device [face mask], it took a little bit longer than I needed it to take to activate. There's a little chemical reaction that takes place inside the canister and I immediately had to take that mask off, grab another one and activate it ... I did not inhale anything and I don't think anyone else did, because ... everyone immediately went to the oxygen ventilators.*

Film released by mission control showed some crew wearing full-face, rubber gas masks while others had masks covering only their mouths and noses. Three fire extinguishers, each containing 3½ pints of a water-based foam, were emptied before the fire was put out. The task was complicated by the way zero gravity caused the foam to surround the flames while leaving a 'bubble' of oxygen in the centre on which the fire could feed.

The minutes during which the fire blazed seemed a lifetime for the crew. Even when it was extinguished, they were shrouded in smoke for some 5 minutes. They were obliged to dismantle the airline and retire to their Soyuz craft while ground control considered whether they should abandon the station.

Eventually it was decided to soldier on by turning on all ventilation systems. However, the crew had to wear gas masks for another two days. Physician Linenger carried out a series of medical checks, looking for possible inhalation damage to the lungs and measuring oxygen saturation in the blood. He concluded that his negative results were because of the rapidity with which the crew were able to don their masks.

## Spare Time

Most space missions are not noted for long periods of idleness, but crews are allowed free time. The ways they choose to pass this time are many and varied, but there is one universal favourite – watching Earth.

'Looking back at the Earth was one pastime I never got tired of,' recalled Ken Mattingly. 'One thing every spacecraft ought to have is a big bay window to sit in and be able to appreciate the view,' Stuart Roosa agreed.

Only in space is it possible to watch the ever-changing face of the planet and stare in wonder at the magnificence of Nature's creation – forest fires, lightning bolts, towering cumulonimbus clouds, jet streams marked by white streamers of condensed vapour, swirling ocean currents, spiralling storm systems, snow-capped peaks, shimmering auroral curtains.

In contrast to this natural splendour, human artefacts are rarely seen. Political borders

*The celestial rainbow of an orbital sunset as seen by the crew of STS-39.*

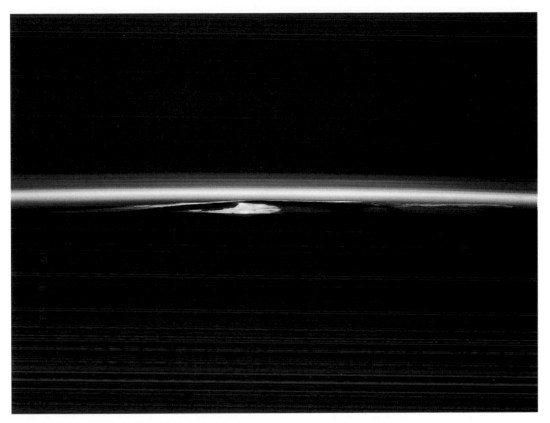

are almost impossible to pick out, unless delineated by some subtle variation in vegetation or land use. The most obvious sign of intelligent life emanates from cities and man-made fires, which glow in the darkness of the night passage. Circular patches in desert regions surround irrigation wells. Otherwise, mankind's creations tend to blend into the background, highlighted only by the occasional glint from a glass structure or a railway track, or the white wake from a solitary ship.

Sometimes, it is human destruction rather than human creation that catches the eye of a space traveller. Estuaries marred by brown eddies reveal regions of deforestation and rapid soil erosion. Salt flats around the Aral Sea and Lake Chad show shrinking waters and desiccation related to climate change, over-irrigation and poor custodianship of the semi-desert environment.

This capacity of the human eye to pick out minuscule features from an altitude of more than 100 miles came as a surprise. Many people refused to accept the incredibly detailed eyewitness accounts of Gordon Cooper in 1963, when manned spaceflight was in its infancy:

*I could detect individual houses and streets in the low humidity and cloudless areas such as the Himalayan mountain area, the Tibetan plain and the southwestern desert area of the US. I saw several individual houses with smoke coming from the chimneys in the high country around the Himalayas. The wind was apparently quite brisk and out of the south.*

*I could see fields, roads, streams, lakes. I saw what I took to be a vehicle along a road … . I could see the dust blowing off the road, then could see the road clearly … . I saw a steam locomotive by seeing the smoke first; then I noted the object moving along what was apparently a track. … I also saw the wake of a boat in a large river in the Burma-India area … and a bright orange light from the British oil refinery to the south of the city [of Perth].*

Regrettably, the last 30 years of human activity seem to be causing a deterioration in the clarity of this orbital window on Earth as smoke and other pollutants pour into the skies.

Another common theme among all space travellers is the unity of Earth as a planet, home to the entire human race. 'You get to thinking of the Earth as an entity,' said Stuart Roosa. 'You don't get to thinking of it as Texas or the United States. You really think of it as Earth.' This feeling of warmth and a sense of belonging was particularly strong for the Apollo astronauts, who saw the little blue dot in the sky as 'a grand oasis in the vastness of space'.

On missions that fly close to the poles, auroras entertain the watching crew. But perhaps the most spectacular display of all is presented by the appearance and disappearance of the Sun once every 45 minutes. John Glenn was the first American to describe the wonder of these celestial scenes:

*Orbital sunset is tremendous ... a truly beautiful, beautiful sight. The speed at which the Sun goes down is remarkable. The white line of the horizon, sandwiched between the black sky and dark Earth, is extremely bright as the Sun sets. As the Sun goes down a little bit more, the bottom layer becomes orange, and it fades into red and finally off into blues and black as you look farther up into space.*

For those few who tire of watching the terrestrial panorama unfold through the porthole, more traditional entertainments such as reading and music are available. Each morning, the astronauts awake to reveille in the form of music, often selected by mission control as personal favourites of the crew. Taped music and personal stereos are commonplace. Occasionally, astronauts with a particular musical talent take their instruments aloft to entertain their colleagues.

## WEIGHTLESS DREAMS

It may be pleasant not to have hard objects protruding into one's flesh, but it takes a while to get used to the idea. Sleeping is a case in point. After years of lying in bed and feeling a mattress beneath one's back, it comes as a surprise to lie down on nothing but air. 'The first night was miserable for me because I felt like I was going to get seasick,' said *Apollo 16*'s Charlie Duke. 'You close your eyes and your head doesn't want to go anywhere, it just sorta stays there. I kept wondering, "Where's my pillow? Where's my head?" So to feel some pressure, I wedged my head up under the couch and the strut, but then I floated loose.'

Zero G allows astronauts to sleep more or less anywhere – it doesn't really matter whether you choose the floor, the wall or the ceiling and whether you choose a horizontal, vertical or inverted position. However, there are certain constraints. Free floating is out, because the unconscious crew member is likely to be awakened by periodic collisions with lockers or instrument panels. Flailing arms and legs during disturbed dreams can cause the slumbering astronaut to damage both himself and the spacecraft. It is, therefore, essential to attach the sleeping bag to something solid and tie down any loose limbs or zip them inside.

Shuttle astronauts can choose to sleep in their seats, in sleeping bags, in bunks (if these are provided) or simply by tethering themselves to the orbiter's walls. Four bunks are installed on the mid-deck for use during missions when the crew is split into two work shifts. This allows one shift to sleep undisturbed by their colleagues. Each 'bed' consists of a padded board with a fireproof sleeping bag attached to it and perforations for ventilation.

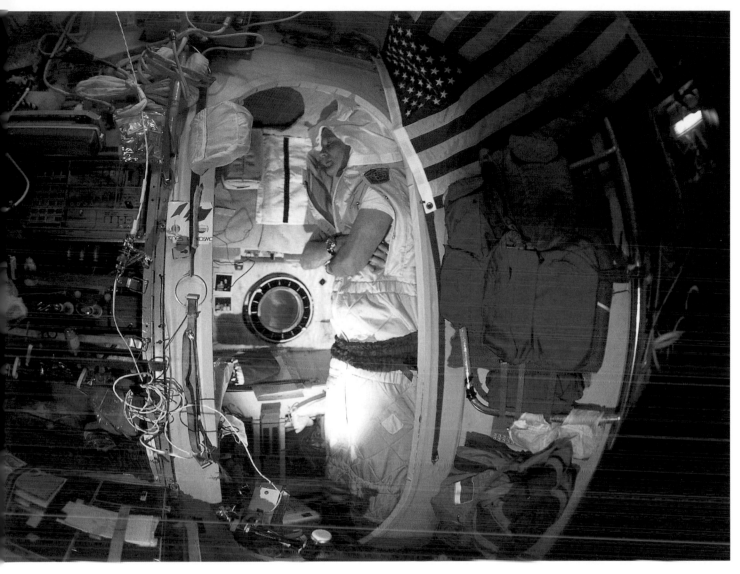

Norman Thagard strapped down for a good night's sleep in one of the private cubicles on the Mir station.

Since the Sun rises and sets once every 90 minutes in low-Earth orbit, long periods of darkness are usually not available. It may be possible to shut out the sunlight by pulling down the window blinds, but many astronauts also take the precaution of wearing blind-folds to ensure uninterrupted sleep.

Having found a comfortable, relatively noise-free spot, the astronaut must take care to check out the ventilation. Exhaled carbon dioxide tends to collect around the sleeper's head, eventually asphyxiating him, so a steady air flow is essential to blow away the deadly gas.

Sleep patterns may well be disturbed by the continual shift from day to night leading to stress or depression. Mission controllers try to avoid upsets to the crew's body circadian clocks by using mission elapsed time, so that the crew sleep when it is night in Houston.

As the crew adjust, dreams often reflect orbital experiences. During his long stay aboard the *Mir* space station, David Wolf reported: 'I'm definitely dreaming in Russian and the people float in my dreams, and that happened after about six weeks.'

Compared with the relatively organized, uncluttered interior of a shuttle, waking up in a space station is indeed an experience in itself. Here is Wolf's account of a typical wake-up session.

*I awake against the ceiling of a densely packed storage area of the* Kristall *module ... . It's the place where I've been temporarily sleeping while spacewalk activities are underway in my usual 'cabin', the* Kvant *back-up airlock. Pushed a space-shuttle-delivered water bag away from my face. Fumbled in the blackness of the night side for that spot of Velcro holding my mini-Maglite and Sony Discman. Faintly heard it still repeating 'Dark Side of the Moon'. Floated out of the marginally tethered sleeping bag and banged my head on the helmet of a ragged old spacesuit, long since cannibalized for parts. Cranked open the micrometeorite cover of the heavy quartz window and, wow, there's Earth.*

## The Heavens and the Earth

Spaceflight has had a profound psychological effect on some astronauts. Some dwelled on the way their surroundings affected their ideas of mankind's place in the Universe. *Skylab* astronaut Edward Gibson commented on the possibility of extra-terrestrial life. 'Being up here and being able to see the stars and look back at the Earth and see your own Sun as a star makes you realize the Universe is quite big, and just the number of possible combinations which can create life enters your mind and makes it seem much more likely.'

Personal relationships were also re-evaluated. 'I now have a new orientation of almost a spiritual nature,' said another *Skylab* occupant, Bill Pogue. 'My attitude toward life and toward my family is going to change. When I see people, I try to see them as operating human beings and try to fit myself into a human situation instead of trying to operate like a machine.'

Others have undergone a more profound spiritual transformation. After their return home from the Moon, *Apollo 15* astronauts Jim Irwin and Alfred Worden set up a non-profit-making Christian organization called the High Flight Foundation to share their faith in God.

Religion often played a part in the thoughts and speeches of the Apollo crews in particular. During their Christmas flight around the Moon, the crew of *Apollo 8* read the first five verses from the Book of Genesis, and commander Frank Borman, a lay preacher in the Episcopal Church, recited a prayer for his congregation. Similarly, Tom Stafford, who commanded *Apollo 10*, requested his pastor to read out a prayer: 'Oh Lord, our Lord, how excellent is thy name in all the Earth, who hast set thy glory above the heavens ... .'

Not everyone appreciated these public displays of religious devotion, and atheist Madalyn Murray O'Hair took out a lawsuit over the *Apollo 8* crew's reading of Genesis. This led to increased sensitivity on the subject by NASA, a sensitivity that was reflected in the secret decision to allow Buzz Aldrin to take a wafer, a tiny gold chalice and a thimble of red wine to celebrate the first Holy Communion on another world.

# 4
# DINING OUT

Whenever explorers have set out to discover new lands, one of the major problems has always been the supply of fresh water and food. Although there are no ports of call on an Earth-orbiting spaceship, the same difficulty arises. On long journeys, fresh supplies of food and drink must be obtained at regular intervals. Not only must an astronaut's daily energy requirement of around 2700 calories be met, but the necessary vitamins, proteins and minerals must be available to provide a balanced diet.

Ensuring sufficient nourishment was not the only problem facing early mission planners. In the early 1950s, before people ventured into space, some doctors and scientists doubted whether astronauts would be able to digest their food or even swallow it without choking.

## DRINK

Anyone who has carried a bucket of water will realize that this most precious of liquids is quite heavy, so the cost of hauling it into space is high. In the 1950s it was estimated that each astronaut would require a daily intake of about 4 pints (around $4^1/2$ pounds) of water, so this amount was loaded onto the short Mercury flights. It was later realized that many space-travellers were losing weight through dehydration, particularly as a result of increased kidney activity, which meant they urinated more frequently. It is now recommended that crews drink at least 5 pints a day, and double that amount if space-walks are scheduled.

*Acting chef Thomas Jones cuts open a food package on Endeavour's mid-deck. Ranged behind him are the food trays belonging to each crew member.*

The first Russian crews drank from a polythene container fitted with a pipe and a cartridge, which both killed bacteria and deodorized the water. Gagarin watched in fascination as droplets that escaped from the mouthpiece floated around his cabin. On Mercury the water was obtained straight from the Cocoa Beach public water system, with no additives, and stored in a small flexible pouch. An astronaut simply squeezed the bag and sucked from a valved tube.

These primitive methods were replaced by a dispenser gun for the two-man Gemini series. Supplies for up to two weeks were stored in the re-entry module and adapter section. To ensure disinfection and prevent microbial growth, the ground-loaded water was treated with chlorine.

With the introduction of fuel cells on US spaceships from the mid-1960s, astronauts were almost swimming in water. By combining hydrogen and oxygen to create electricity, the cells produced water as a by-product, generating a gallon every $7^1/_2$ hours. It was originally intended to use the fuel cell water for consumption, but its unappetizing taste, odour and colour meant that all Gemini missions relied on ground supplies.

The fuel cells were redesigned for the Apollo command module to generate potable water on-board, but in many ways the system was similar to that in Gemini. Excess water was dumped overboard from a waste tank. To prevent contamination, a sodium hypochlorite solution was injected daily by the astronauts at a port upstream of the storage tank.

Fuel cells are still used on the space shuttle. Excess hydrogen is removed from the water and dumped overboard. The water is then routed to any of four 20-gallon tanks, although normally only two of these are used. Iodine is automatically added to the water from a small canister installed in the system.

Such sophisticated technology is not available on the small Russian Soyuz craft that ferry a crew of three to and from the *Mir* station. This is relatively unimportant, however, because rendezvous and docking take only two days, and re-entry is usually completed in a matter of hours. The craft carries just over 5 gallons of cold water.

Supplying space station crews over many months is more of a challenge. For *Skylab* the Americans decided to launch everything that was needed at the beginning of the mission, since the absence of fuel cells meant that no on-board water generation was possible. The available water could hardly be said to be 'fresh', for the ten 72-gallon capacity stainless steel tanks located in a ring around *Skylab*'s outer skin had to last the lifetime of the station. Each tank was dedicated to a specific application, and at any time, only one of them was used for human consumption, with a second for hygiene. Levels of iodine were continually monitored, particularly after long periods of stagnation between flights, and more was injected when necessary by the resident crew.

The Russians adopted a different approach. The *Mir* station is equipped with four 55-gallon tanks, which can be partially replenished every few months by visiting Progress cargo ships. Until 1980, the water was shipped up in 1.3-gallon containers, which had to be hauled aboard by the resident Salyut crew. Since the arrival of the *Progress 9* cargo ship in 1980, the cosmonauts have been able automatically to pump 47 gallons aboard. Additional water has been delivered by visiting shuttles over the past few years.

The crews are not completely dependent on such visits, however. Russian space stations are equipped to recycle urine and condense moisture from the air. Every 2 pints of urine yield about $1^1/_2$ pints of distilled water, although, for psychological reasons, it is not used for human consumption but undergoes a process of electrolysis that splits it into hydrogen and oxygen for breathing.

## THE SOFT DRINKS WAR

A crew would soon be up in arms if all it had to drink was water, and since the early days of space travel astronauts have been able to sip fruit juices and milk. Later, with the advent of hot water, the range of beverages was expanded to include cocoa, tea with or without lemon, and coffee with cream, as well as there being a choice between sugar and artificial sweetener.

The drinks themselves were fine, but the gas bubbles in the water and additives recommended by medical experts tended to cause problems. On one occasion astronauts on the Moon inadvertently treated listeners to a catalogue of their digestive problems. The culprit was identified as an extra dose of potassium in the orange juice, which had been added in an attempt to stabilize their heart function:

*I haven't eaten this much citrus fruit in 20 years. But I'll tell you one thing, in another 12 f-----g days I ain't never eating any more. ... I like an occasional orange, I really do, but I'll be damned if I'm going to be buried in oranges.*

Coffee is about the strongest stimulant available to astronauts, since NASA maintains a teetotal policy, but the Russians are not quite as strict. One of the few recorded instances of alcohol being imbibed in space came in April 1979 when *Soyuz 33* failed to dock with the *Salyut 6* space station. Bulgarian Georgi Ivanov attempted to raise the spirits of his Soviet companion, Nikolai Rukavishnikov, by opening a gift-wrapped bottle intended to be drunk during the post link-up party. 'I had very little but Georgi took a good drink,' commented Rukavishnikov.

A similarly stressful situation in 1997 caused cosmonauts Lazutkin and Tsibliev to steady their nerves with a nip of cognac after a Progress craft crashed into the *Mir* station. A Russian spokesman explained: 'Officially, alcohol is ... prohibited. Unofficially, they can bring very limited quantities.' He added rather sheepishly, 'Nobody ever gets drunk.'

Meanwhile, soft drinks companies see space as a great place to advertise their products. In the days of Apollo, the manufacturers of Tang Orange announced that theirs was the astronauts' favourite breakfast beverage. No one seems to have asked the astronauts. *Apollo 11* crewman Buzz Aldrin later commented: 'The three of us sampled the orange drink, supposedly Tang, and instead chose a grapefruit-orange mixture as our citrus drink.'

While NASA avoids commercial sponsorship on its missions, Moscow is now only too willing to earn extra dollars for its hard-pressed space programme through advertising. One recent presentation filmed for an Israeli dairy company showed *Mir* commander Vasili Tsibliev chasing floating globules of milk in an effort to swallow them. Another cosmonaut, Alexander Lazutkin, was obliged to extol the virtues of bananas in zero gravity.

The ultimate example of commercial rivalry spreading into space is the competition between soft drinks giants Coca-Cola and Pepsi-Cola. Unfortunately, fizzy drinks soon lose their fizz in zero G as the carbon dioxide separates from the liquid. The cola giants, eager to expand their empires into space, struggled to make a can in which the contents mix before emerging from the drinking spout.

It began in 1985 when shuttle astronauts tried specially adapted cans of Coca-Cola and Pepsi-Cola. NASA diplomatically kept quiet about which brand was preferred by the astronauts. The cola experience was repeated six years later on *Mir* by cosmonauts Artsebarski and Krikalev.

Competition came to a peak in the spring of 1996. First, Russian cosmonauts Yuri Onufrienko and Yuri Usachev posed for the cameras during two spacewalks, holding up an inflatable Pepsi can and a sign that read: 'Even in space, Pepsi is changing the script.' An advertising company executive crowed: 'We have definitely taken Pepsi to new heights.'

A more serious approach was put forward by Coca-Cola and NASA. Soda fountain tests flown on the shuttle in February 1995 and May 1996 were presented as a scientific challenge described as 'two-phase fluid handling'. In simple terms, this meant that the astronauts would use the soda fountain to mix water, syrup and carbon dioxide to produce the right amount of fizz. Despite the scientific sell, not everyone agreed that it was worth NASA spending $3 million, topped up with $750,000 from Coca-Cola. Unfortunately, the soda fountain had to be sent back to the drawing board since it was great at producing fizz but not much else.

The eventual aim is to develop refrigeration and liquid dispensing units that will make soft drinks available throughout the Universe. The first step for Coca-Cola is to place its soda fountains on the ISS. 'Our goal is to take Coca-Cola wherever humans are,' commented company researcher Michael Myers.

*Spacewalking cosmonauts hold up an inflated Pepsi can during a spacewalk on 21 May 1996. The US company was said to have paid £1 million (about $1.6 million) to film an advertisement during two spacewalks. Cosmonauts Yuri Onufrienko and Yuri Usachev were also given baseball caps and T-shirts to wear.*

## DINING OUT

In the 1950s food scientists were faced with the challenge of how to provide space food that was uncontaminated and safe to use in zero G. There were two main concerns: crumbs of food contaminating the spacecraft's atmosphere and floating into sensitive instruments, and the health hazard from disease-producing bacteria, viruses and toxins. To overcome these obstacles NASA enlisted the help of the Pillsbury Co., Minneapolis, and the armed forces.

At first food took the form of paste in aluminium tubes or bite-sized cubes coated in gelatine to reduce crumbling. Taste took a back seat to considerations such as nutritional balance, weight, packaging, storage life and easy preparation.

Inevitably, the first space travellers acted as guinea pigs for those who would follow. Although he was in orbit for little more than 90 minutes, Gagarin was obliged to sample some paste food placed in a small container beside his right shoulder. Gherman Titov was able to try more traditional solid foods – some chunks of bread and vitamin-enriched peas. Meals for later Vostok cosmonauts progressed to packed slices and cubes, spiced up with the odd sausage and tubes of milk or fruit juice.

Neither Alan Shepard nor Gus Grissom had time to eat anything on their short sub-orbital lobs, but John Glenn was obliged to squeeze a tube of apple sauce into his mouth. He remarked sourly: 'I wish now I had brought along that ham sandwich someone once put in the ditty bag.'

It may have been fun to chase and consume floating food, but anything that was wet or disintegrated into crumbs was a potential hazard and banned from orbit. The first major breach of this rule came when Grissom spread a shower of rye bread debris around the cabin from a corned beef sandwich provided by John Young. Mission controllers were not amused, and both men were reprimanded. However, as NASA's senior astronaut, Young later cocked a snook at officialdom by giving Bob Crippen a non-regulation corned beef sandwich on the shuttle's maiden flight.

As time went by, aluminium tubes were abandoned in favour of plastic containers into which cold water was injected through the nozzle of the water gun. After kneading, the contents were squeezed into the astronaut's mouth through a tube.

A wider choice of dishes was also added to the menu, although progress was slow for Russian crews. With no method of reconstituting dehydrated food, cosmonauts on the Soyuz flights were restricted to tubes of purée, dried cubes and tinned meat, including steak, ham, chicken and veal. The main advance came from *Soyuz 9* onwards, when a heater was added for soups and drinks.

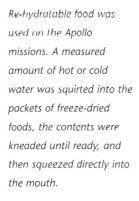

*Re-hydratable food was used on the Apollo missions. A measured amount of hot or cold water was squirted into the packets of freeze-dried foods, the contents were kneaded until ready, and then squeezed directly into the mouth.*

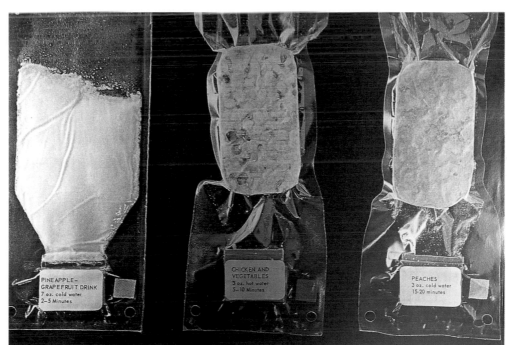

In contrast, American astronauts were fairly pampered. Each Gemini astronaut could choose a combination of food items as long as they provided 2800 calories a day and a balanced diet. A typical cold meal would include shrimp cocktail, chicken and vegetables, toast squares, butterscotch pudding and apple juice.

Opinions about the fare varied considerably. Frank Borman, commander of *Gemini 7*, wrote about: '... the boredom of eating monotonous, tasteless food. The worst items were the beef and egg bites – terribly dry, leaving a bad taste in the mouth and a coating on the tongue. The desserts were too sweet and we would have preferred larger servings of fruit juices.' *Gemini 10* astronaut Michael Collins, on the other hand, seemed in seventh heaven with his packet of cold, rehydrated cream of chicken soup. 'It's the best thing I have ever tasted, better than a martini at Sardi's, better than the pressed duck at the Tour d'Argent,' he lyricized.

Astronauts now take for granted the availability of a wide choice of hot meals and drinks, but such luxuries only became available with the introduction of fuel cells on the Apollo spacecraft, which made heated water possible. Further advances in Apollo facilities came with the introduction of rehydrated foods stored in a spoon-bowl. When the zipper was opened, the moisture content was sufficient to cause them to cling to a spoon.

Unfortunately, there was another side to the use of water to rehydrate food – flatulence. Buzz Aldrin described this unfortunate side effect:

*It wasn't long before we discovered that the little device designed at the last minute to ventilate hydrogen from the water, as it passed from the gun to the food bag, was not always the success its designers had hoped. Frequently, the hydrogen ... stayed in the water and was swallowed by us. The result was stomach gas. At one point on the trip back to Earth it got so bad it was suggested we shut down our attitude control thrusters and do the job ourselves.*

Despite the gas bubbles, dining conditions became more civilized on *Skylab* and the Soviet space stations. For the first time there was sufficient room for a dining table. Crews could 'sit down' at the table by means of foot and thigh restraints. Unfortunately, *Skylab* crews were notoriously hard to please and criticized the table mercilessly.

Mission planners were able to take advantage of the large storage space on *Skylab* by providing a choice of 72 food items. There were also a freezer and a refrigerator. Daily menus were pre-selected before launch, however, and the fact that these were repeated every six days led to dissatisfaction. A despairing Jack Lousma wailed: 'I've asked myself every six days, whenever it turns up on the menu, "How come I picked beef hash for breakfast?"'

The food trays were another source of complaint. They were designed both to hold the meal in place and serve as heating devices, and underneath three of the eight cavities in the trays were warmers that were meant to raise temperatures of particular dishes to 66°C. In practice, they didn't work well and many tepid meals were consumed.

*Skylab* food containers consisted of aluminium cans with pull-out lids. Rehydratable foods were stored in a plastic pouch inside the can and had a water valve for rehydration. Ready-to-eat foods were held in their cans with a slit plastic cover. Generally the precautions proved successful, although a piece of meat or other food would occasionally detach itself and float away.

Instead of plastic bags, *Skylab* drink containers were collapsible bottles that expanded when filled with water. Eating on *Skylab* was fairly normal – knife, fork and spoon were held magnetically to the food tray until needed, although they tended to float away once the cans were opened. Scissors were added to enable the plastic covers to be cut open. Once again, the astronauts criticized their utensils. The cutlery was too short to reach the bottom of the bags, while the spoon bowls were too small for man-sized mouthfuls.

Although the space shuttle does not have a freezer or refrigerator, it does benefit from the addition of a galley on the mid-deck. This features hot and cold water dispensers, a pantry, a convection oven, individual serving trays and a water heater. A hand-washing hygiene station is attached to the galley's side.

Three one-hour meal periods are scheduled for each day of the mission. This hour includes actual eating time and the time required to clean up. Breakfast, lunch and dinner are scheduled as close to the usual hours as possible, and dinner is eaten at least 2–3 hours before the crew's sleep period. Food preparation begins 30–60 minutes before mealtime and is usually taken in turns by members of the crew. Heating and reconstitution of the food takes an additional 20–30 minutes.

Crews choose their menus about 5 months before lift-off. Each astronaut's food package is identified by a coloured dot and has a Velcro tab on its base to hold it in place on the food tray. Meals are stored in locker trays with food packages arranged in the order they will be used. Dining tables are out; instead, astronauts eat from trays tied to their laps or fixed to a wall.

Hot or cold water in pre-measured amounts is injected into rehydratable packages by inserting a large hollow needle through their polyethylene base. An identical design for drink containers means that a straw can simply be inserted through the same septum used for injecting water. Pouches or plastic containers of food requiring heating are placed in the forced air convection oven, which has a maximum temperature of 82°C.

The shuttle menu includes more than 70 food items, as well as serving-sized packets of condiments, and 20 beverages. The fresh food locker contains bread, breakfast rolls and fruit and vegetables, such as apples, bananas, oranges, carrots and celery. Pantry items are also available to provide extra drinks or snacks and as a precaution against the unexpected extension of the mission.

Following the meal, the utensils and trays are wiped clean at the hygiene station with pre-moistened towelettes.

Eating while wearing a space suit is obviously not possible. Spacewalking is, however, tiring, thirsty work, so astronauts carry a drinks bag and an 'energy bar', attached by Velcro inside their helmets. Not surprisingly, mishaps happen from time to time. At the beginning of his first Moonwalk, *Apollo 16* astronaut Charlie Duke's enthusiasm was dampened somewhat by a leaking bag of orange juice. 'I've already had an orange shampoo with the helmet on. I wouldn't give you two cents for orange juice as a hair tonic.'

On Russian space stations, most of the food is packaged in tins and tubes and can be heated to 65°C for 10–20 minutes in a small oven in the dining table. For deep-frozen food, the temperature is raised to 75°C. A refrigerator and freezer are installed on *Mir*, although their maintenance has caused some problems and both have ceased to work at various times.

For American visitors, Russian space cuisine proved quite a culture shock, although opinions varied. Norman Thagard had little good to say about the fare on *Mir*, but Shannon Lucid declared cabbage and beef stew to be her favourite, while Andy Thomas particularly liked the soups.

From time to time, foreign 'guests' on the Russian stations bring unusual, and sometimes exotic, cuisine. In 1984 Indian Rakesh Sharma introduced his cosmonaut

*Cosmonauts Alexander Kaleri and Valeri Korzun (centre) appreciate the delivery of fresh oranges and grapefruit by the STS-81 crew.*

colleagues to curry, crisp bananas, mango fruit bars, and pineapple and mango juice. However, first prize for orbital cuisine must go to the French. The dishes accompanying Jean-Loup Chrétien on his first trip into space included jugged hare, crab soup, country pâté, lobster pilaff rice with sauce à l'armoricaine and Cantal cheese. A gourmet menu for Leopold Eyharts included quails stuffed with grapes, boeuf en daube en Madiran and tomatoes à la provençale.

One of the most notable features of long missions is the deterioration of the cosmonauts' appetite and sense of taste. Some cosmonauts develop crazes for particular foods. During a 140-day marathon on board *Salyut 6*, Alexander Ivanchenkov became obsessed with cheese, even coveting his partner's rations. Shannon Lucid developed a craze for M&M sweets.

Congested heads and nasal passages often destroy the senses of smell and taste, causing even favourite foods to resemble sawdust in zero G. On *Skylab* the crews resorted to large doses of salt to spice up their bland fare. Not surprisingly, Russian crews look forward to the arrival of Progress supply craft loaded with items such as onions and garlic, and fresh foods such as fruit, bread and milk.

Financial cutbacks since the breakup of the Soviet Union have placed these extras in jeopardy. In January 1992 cosmonaut Dr Valeri Polyakov complained about difficulties in obtaining certain foodstuffs. 'We used to get honey from the republics, but now they have quit deliveries,' he complained. Poliakov expressed the fear that these changes could reduce vitamin intake and affect the rate of cosmonauts' rehabilitation after landing.

Attempts to augment the cosmonauts' diet by growing vegetables in a 'space garden' have met with limited success. Most crops have stubbornly refused to grow properly or produce seeds, and experiments show that the plants thrive best when grown in a centrifuge that creates artificial gravity. Among the crops planted and carefully tended by Salyut and *Mir* crews are garlic, onions, cucumbers, wheat, lettuce, peas, parsley and dill.

One unusual experiment carried out on board *Mir* involved the use of an incubator to rear quail chicks. On the first occasion, eight out of 48 eggs hatched, but the young birds were unable to adapt to zero gravity and they all perished within a few days. A second attempt in November 1992 was equally unsuccessful. Scientists hope that these problems can one day be overcome, allowing cosmonauts to tuck into poultry meat and fresh eggs on their journeys through the cosmos.

## WASTE DISPOSAL

One of the major problems facing technological countries is how to dispose of all the waste people generate. This is a particular problem in orbit, where every nook and cranny is crammed with the paraphernalia required for human spaceflight. Much of the waste is meal-related and results from lack of washing-up facilities and the need for bulky packaging.

The crew of *Gemini 7* was one of the first to come up against the waste-disposal problem during the two-week mission in 1965. Jim Lovell described their solution:

*We were worried that we'd sort of get pushed out of the spacecraft with all the debris that we would accumulate. So we spent many hours prior to the flight finding little spots and crevices in the spacecraft where we could pack things. We would eat three meals a day, and Frank [Borman] would very nicely pack the containers in a small bag, and at the end of the day he would throw it behind the seat. We managed to get nine days' debris behind those seats.*

The Apollo astronauts were less fastidious. Once they had eaten, the used meal bags were dosed with tablets that dissolved the food residue, then rolled up tight, deposited into a plastic container and squeezed into storage lockers. When these began to overflow, the crew dumped them overboard, where they became miniature satellites.

Similar methods have been used on space stations such as Salyut and *Mir*, although the Russians also have the convenience of their own space incinerator. Every few months an automated Progress ship brings several tons of fresh supplies, and once it is unloaded, the cosmonauts fill the ferry with garbage and return it into the atmosphere, where it burns up during re-entry.

No such system was available on *Skylab*. William Pogue explained the procedure:

*Because we ate directly from plastic bags or cans, the only things that required cleaning were our tableware and food trays. These were wiped with tissues soaked with a mild disinfectant. The cans were crunched flat with a special food-can crusher and placed in a bag for disposal.*

A similar system is used on board the shuttle. However, while trash on the shuttle is returned to Earth, *Skylab* was so large that it had a built-in rubbish dump in the form of a 2000-cubic-feet tank. The *Skylab* astronauts pushed their waste (including soiled clothes and urine bags) through a tube fitted with an airlock to prevent loss of air when they opened the hatch. This worked well at first, but caused increasing problems as time went by.

Pogue described his crew's solution:

*We finally worked out a system whereby Jerry Carr would load the trash bag in the bin of the trash airlock and I would float above, holding onto the ceiling. As he closed the hatch, I would pull myself down sharply and stomp on the hatch lid while Jerry closed the locking lever.*

There was no need to worry about how to empty the tank and dispose of the garbage. Nature eventually accomplished that task when friction with the upper atmosphere caused a gradual lowering of *Skylab*'s orbit until it fell to Earth and was incinerated.

## The Human Vacuum Cleaner

One of the most frequent questions put to astronauts is: 'How do you go to the bathroom in space?' The answer depends on who you ask.

For the first American space shot, the question does not seem to have occurred to anyone. Alan Shepard's suborbital lob from Florida into the Atlantic would take only 15 minutes so the provision of toilet facilities was considered unnecessary. Unfortunately, the launch was delayed by technical hitches, and as the hours ticked by, Shepard became increasingly uncomfortable, until he could hold back no longer. Informed that it was impossible to leave the capsule, the agonized astronaut told capsule communicator Gordon Cooper that he was 'going to let it go in my suit'. There followed one of the most unusual conversations of the entire space era.

*Cooper: No! Good God, you can't do that. The medics say you'll short circuit all their medical leads!*
*Shepard: Tell 'em to turn the power off!*
*Cooper (after a hurried conference): OK, Alan. Power's off. Go to it.*

As luck would have it, Shepard's reclining position allowed the urine to pool in the hollow of his back, where the thick underwear soaked it up. Cool oxygen flowing through the suit acted like a breeze on a washing line, allowing Shepard to concentrate on the mission ahead.

Unwilling to endure the same embarrassment and discomfort, America's second man in space, Gus Grissom, told NASA to find a solution. Reasoning that any tight-fitting garment would serve as a temporary stopgap, flight surgeon Bill Douglas sent nurse Dee O'Hara into Cocoa Beach to buy a panty girdle. So it was that Grissom risked both his neck and his reputation by launching into space wearing an item of women's underwear.

A more permanent answer had to be found. After testing many designs, astronauts were provided with a simple bag with a condom-like bladder worn around the waist. The urine collected in the nylon bag could be emptied directly into space through a hose linked to a dump valve on the right thigh of the pressure suit. While free of the suit, the astronaut could dump his urine directly overboard. The only urine stored on board was for medical analysis back on Earth.

Solid waste was a different proposition. Until the space stations came along, going to the toilet was hardly a high-tech process. As Gemini astronaut Mike Collins commented: 'It was bad enough to have to unzip your pressure suit, stick a plastic bag on your bottom, and defecate – with ugly old John Young sitting six inches away.' The faecal collection bag was about 8 inches across, with an adhesive ring around the top that was pressed against one's flesh. After use, chemical tablets were added to kill the bacteria, and the bag was stored in a waste container.

This was fine in theory, but far from foolproof in practice, as *Apollo 16*'s Charlie Duke attested. 'The only thing is that nothing goes to the bottom of the bag in zero gravity. Everything floats. In some cases with a loose [bowel] movement, it was really unmanageable. It just didn't work.'

In the absence of toilet facilities, spacewalkers have to wear a faecal containment

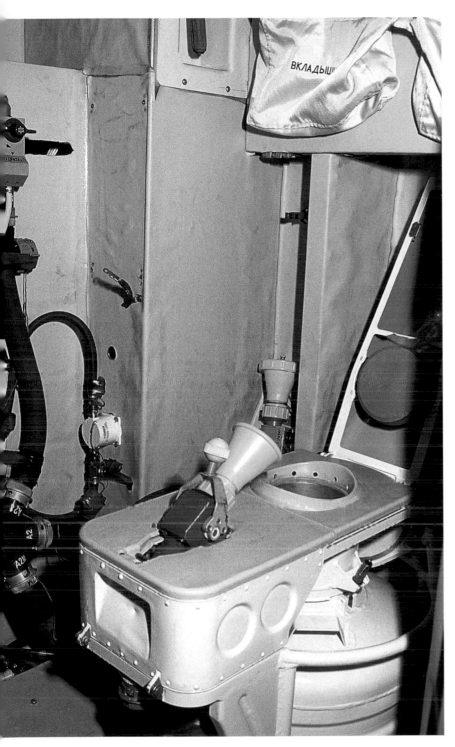

*The toilet on board the Russian Mir station operates with an air-suction system. Urine is recycled to make pure water and oxygen.*

garment, a pair of tight-fitting fabric shorts, which work rather like a disposable nappy. Once the EVA is over, any faeces are transferred into the appropriate containers. Men wear a urine collection device with a rubber sleeve, which is attached by Velcro to their underwear.

The advent of women astronauts meant a rethink over urine disposal. While wearing a spacesuit, women astronauts resort to commercial adult diapers or use a similar garment called a disposable absorbent containment trunk, which is capable of soaking up 2 pints of liquid.

Things became rather more civilized with the arrival of space stations. As spacecraft became roomier and flights began to last for many months, proper toilet facilities became a necessity.

The basic principle behind all space toilets is the vacuum cleaner. On *Skylab* the astronaut used the available restraints to hold himself in position, then took advantage of either the funnel-shaped urine collection device or the commode. Air drawn through the toilet provided suction. Urine was collected in a bag, which was changed daily. Solid waste was also collected in a bag, which was changed after each use. The faeces were dried in a heat-vacuum chamber and stored for examination back on the ground.

The principle is similar on the shuttle and the Russian space stations, although for mixed shuttle crews the urine collection device had to be altered so that it became unisex. However, even multi-million dollar toilets sometimes come to grief. On these occasions, the crews have to resort to the old, tried-and-tested methods of bygone days. As one shuttle astronaut commented: 'We decided that those Apollo astronauts must have been real men.'

The original shuttle toilet had so many shut-downs that NASA felt obliged to spend more than a million dollars on improvements. Needless to say, critics had a field day with the astronauts' 'gold-plated' lavatory. The improved version, which first flew on STS-54, was designed to be more dependable, more hygienic and did not need to be removed for maintenance between flights. It is equipped with a waste compactor and has almost unlimited capacity.

# 5

# THE SPACE TRANSPORTATION SYSTEM

The modern era of space transportation began on 12 April 1981, the twentieth anniversary of Gagarin's pioneering flight. As space shuttle *Columbia* blasted off from Florida on her maiden voyage, reusable spacecraft became a reality. It was the first step on the road to making spaceflight safe and available to all.

Since then, almost 100 space shuttle flights have taken place, but the excitement and sense of adventure have largely dissipated. Most missions now start and finish with barely a mention from the world's media, and most people are not even aware that, every six weeks or so, half a dozen individuals spend a week or more in orbit, 200 miles above our planet. Flights grab the headlines only when lives are in danger, expensive hardware breaks down or woodpeckers drill holes in high-tech insulation on the fuel tank.

## THE SHUTTLE FLEET

The space shuttle concept was born before Armstrong and Aldrin's historic steps on the Moon. Originally conceived as a ferry craft for a space station, the 1969 designs revolved around a two-stage, fully reusable vehicle.

It soon became clear that this would not be a viable prospect given the budget cuts introduced by President Nixon. NASA scaled down its specifications, and by the end of 1971 it was looking at a delta-winged manned orbiter, an expendable external fuel tank and two recoverable solid fuel boosters. Development costs were estimated at $5.2 billion, with each flight costing $10.5 million. After a maiden launch in 1978, the flight rate would eventually build up to 50 flights a year.

In the agency's early publicity material NASA waxed lyrical about the capabilities of the new vehicle:

*Four shuttle orbiters, each with accommodations for up to seven persons, are scheduled to make as many as 60 round trips a year to low Earth orbit and back. Shaped like an airplane, the 37.2-metre [122-foot] long shuttle orbiter will lift off like a rocket, orbit like a spacecraft on seven to 30-day missions and return to Earth on a landing strip like a glider or airplane.*

The agency went on to tell the world of its multi-functional capability, which would revolutionize space travel. It would deliver and launch satellites in space. It would serve as an orbital repair shop for ailing satellites. It would operate as a science lab. It would become a factory, manufacturing new products, which could be made only in zero gravity. And it would deliver supplies and assist in the construction of future space stations.

Although the predictions of its flight efficiency and duration proved hopelessly optimistic, the shuttle has fulfilled all the roles its designers envisaged. In addition, instead of remaining the exclusive preserve of superfit test pilots, near-Earth space has been opened to men and women aged anywhere from their early 20s to their late 70s.

Six shuttles have been built, although only five have flown in space. *Enterprise*, named after the starship in the TV series *Star Trek*, was a test vehicle and never intended to fly outside the atmosphere. The first operational shuttle was named *Columbia* after a US Navy frigate that circumnavigated the globe in the early nineteenth century. It is nearly 4 tons heavier than the other orbiters, so its payload lift performance is reduced by a similar mass. The others weigh in at around 173,000 pounds (77 tons) dry weight, and are capable of lifting 26.6 tons into a low-inclination orbit or 20.3 tons into an orbit inclined at 57 degrees to the equator. *Challenger* was named after a nineteenth-century ocean exploration ship. Its first launch came in 1983, but after nine successful missions it was destroyed in a launch explosion on 28 January 1986.

*Discovery* came along in 1984. It was named after two famous sailing ships: the vessel in which Henry Hudson tried to find the Northwest Passage and the ship used by Captain Cook to discover Hawaii. The fourth addition to the fleet was *Atlantis*, whose maiden flight took place in 1985. It was named after a sailing ship once operated by the Woods Hole Oceanographic Institute. *Challenger*'s replacement was named *Endeavour* after the first ship commanded by Cook during his exploration of the south Pacific. Its maiden flight began on 7 May 1992.

## A TOUR OF THE SHUTTLE

Shuttles are designed to carry between two and eight astronauts, although 10 people can be squeezed inside in an emergency. Despite its size – the shuttle is 122 feet long and has a wing span of 78 feet – most of the orbiter is taken up with the payload bay. The crew occupy a relatively small pressurized cabin section at the front end.

There are three levels to the crew cabin. Uppermost is the flight deck where the commander and pilot control and monitor the progress of the mission. The commander sits on the left, the pilot on the right. During the launch of a seven-strong crew, two other astronauts are seated towards the rear of the flight deck.

Apart from six large windows at the front of the vehicle, there are two overhead windows towards the rear of the cabin and two aft-viewing windows overlooking the cargo bay. These are particularly valuable for ensuring a safe docking with a space station or for monitoring satellite operations and spacewalks.

As the orbiter's control centre, the flight deck contains an awesome array of separate displays and controls. Altogether, there are more than 2000, three times more than on the Apollo command module. In order to maximize safety, the shuttle is equipped with dual controls so that it can be flown from either the commander's seat or the pilot's seat. If either becomes incapacitated, this arrangement allows a solo emergency return to Earth.

In the centre of the main panel are three computer screens and a range of displays and indicators showing the craft's altitude, acceleration, vertical and horizontal velocity. Between the two seats are the main flight computer, the main engine controls and the autopilot controls. Above their heads are the digital clock showing mission elapsed time, as well as dials and switches for the cabin environment, fuel cells, thrusters and manoeuvring engines. Each also has a control stick between his knees, which can be used for manual manoeuvring. These 'rotational hand controllers' are also used when the shuttle returns to the atmosphere during approach and landing. Pedals operate the rudder for yaw control and the wheel brakes after touchdown.

At the rear of the flight deck are yet more controls arranged in a U-shape in what is known as the aft crew station. The panel behind the pilot's seat is used to monitor orbiter and payload systems, while the one on the right has monitors and controls for each payload.

*Shuttle* Discovery *photographed by an IMAX (wide-screen) camera on board the Orfeus/SPAS satellite in September 1993. The crew cabin at the front occupies about a fifth of the orbiter's total length. The rest is taken up by the payload bay, seen here with the doors open, the thruster section in the nose and the orbital manoeuvring engines by the tail at the rear.*

The main aft-facing console is divided into two sections. With the aid of two control sticks, the pilot 'standing' on the left side can rendezvous and dock with a satellite or space station, while a mission specialist on the right-hand side has another two hand controllers and two TV monitors available during manipulation of the robot arm in the cargo bay. Other delicate operations handled from there include opening and closing payload bay doors, deploying radiators and operating external lights and TV cameras.

Below this cramped work-place is the mid-deck, where the three remaining crew take their launch positions. Astronauts enter this area through two small access hatches, each just a little over 2 feet square. The main entrance and exit hatch is also located on this deck. A ladder on the port side, although superfluous in zero gravity, is required for entry on the launch pad by the astronauts and ground crew.

*Japanese mission specialist Koichi Wakata using the controls of the remote manipulator arm at the rear of the flight deck during STS-72. The large handle in the centre is a hand controller for manoeuvring the orbiter.*

Until recently, access to the airlock was at the rear of the mid-deck in all four shuttles. This small chamber was where astronauts prepared for space-walks and gained access to the cargo bay. Now, only *Columbia* has this arrangement. The other three orbiters, which will be used to dock with the ISS, now have the airlock in the cargo bay. It is reached through a tunnel from the mid-deck. The additional room is particularly valuable since this is where the storage lockers are located, and where the crew eat, sleep, wash and go to the toilet.

A galley may or not be installed, depending on the type of mission and the amount of room required for pay-loads. The oven has a heat range of 60–85°C and is divided into two compartments. In the lower compartment, 14 or more rehydratable food containers can be inserted on tracks, while the upper section can hold at least seven food pouches beneath spring-loaded plates.

Dispensers for liquid salt and pepper and vitamins can be held in place by clips below the rehydration station, and packs of relish are stored in two open-ended boxes attached to the galley's front panel by Velcro tabs. Another nearby container holds individual packets of wet wipes. The shuttle galley was redesigned and upgraded in 1991. The modern version takes up half the volume of its predecessor and (on Earth) weighs one-third less.

In the absence of a galley, individual food trays have to be stored in a mid-deck locker during launch, and they are taken out during preparation for the first meal. The crew has to make do with a food warmer and a water dispenser. The briefcase-sized food warmer contains a hot plate, which operates at 74–79°C. Although it is small enough to be carried around, a meal for a crew of four can be prepared within an hour.

Four sleeping stations may also be installed to one side of the mid-deck when the crew is divided into two shifts. They are arranged so that two people can sleep on the top and lower bunks, while a third person sleeps beneath the lower bunk facing the floor. A fourth crew member sleeps vertically in a bunk set against the end of the two-level bed.

The lower deck is rarely entered, although access is possible through removable panels. Environmental control equipment and additional storage space are found there, and the crew find the extra room invaluable for storing rubbish such as lithium hydroxide canisters and wet towels.

## DAWN OF A NEW ERA

The maiden flight of *Columbia* carried the future hopes and aspirations of a space agency that had not launched a manned spacecraft in almost six years. Yet the dream was already tarnished. Cost overruns and technical problems had delayed the shuttle's maiden flight by two years. By the time *Columbia* lifted off, NASA officials were well aware that their promises of a cheap, multi-purpose space vehicle would not be fulfilled. If the first flight flopped, NASA's entire manned spaceflight programme would be in jeopardy. As it was, the crew, John Young and Bob Crippen, were risking their lives by becoming the first crew to fly on a spacecraft that had never been flight-tested.

On 12 April 1981 the Cocoa Beach area of Florida's Atlantic coast was inundated with motorists and spectators, the largest crowd to watch a space launch for many years. An estimated 50,000 people crammed into the viewing areas, with another 600,000 squeezing into any available space on the beaches, causeways and river banks.

Aboard the orbiter, lying on their backs and surrounded by white vapour venting from the giant fuel tank, the crew waited anxiously for the countdown to reach zero.

This time everything proceeded smoothly. The 'beanie cap' used to vent oxygen was withdrawn from the fuel tank, the crew access arm was pulled back, the auxiliary power units were started, the fuel tank was pressurized and the automatic sequencer took over the count. Launch director George Page read a good luck message from President Reagan. Then came the long-awaited words: 'We have go for main engine start.'

The SSMEs roared into life, but they were drowned 3 seconds later by the ignition of the SRBs. The launch pad was engulfed in a billowing cloud of white steam tinged with orange and yellow. 'We have lift-off of America's first space shuttle!' announced a delighted NASA commentator. At 3 seconds past 7 a.m., America re-entered the space race.

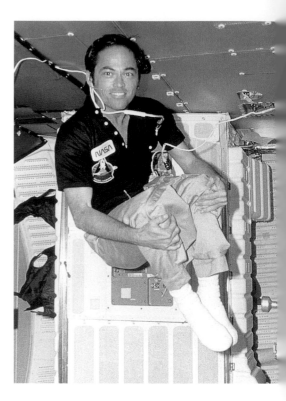

STS-1 pilot Robert Crippen *enjoying the freedom of zero gravity aboard* Columbia. Crippen *transferred to NASA after the military Manned Orbiting Laboratory project was scrapped in 1969. He had been waiting for his first mission for more than 14 years.*

As the shuttle turned upside down and headed eastwards over the Atlantic, the capsule communicator informed the crew: 'You are Go at throttle up.' Soon, they were too high to use their ejector seats. Two minutes 20 seconds into the flight, the SRBs fell away, having completed their work. It was left to the SSMEs to kick the craft the rest of the way into orbit.

As the curvature of Earth appeared in his window, Crippen exclaimed: 'What a view! What a view!' Medical sensors showed that, while his pulse was racing at 130 beats a minute, the pulse of Young, veteran of four previous flights, remained at a cool 80 beats a minute.

Main engine cut off took place at 8 minutes 34 seconds. The giant external tanks, now virtually empty, were released. All that remained was a series of firings with the twin orbital manoeuvring engines to adjust the 185-mile-high orbit.

On the first pass over the United States, the crew opened the cargo bay doors, exposing the cooling radiators to the vacuum of space. The next two days proved to be a fairly routine shake-down cruise, disturbed only by media hysteria over some missing thermal protection tiles on the rear engine pods. The rumour spread that the ship would burn up during re-entry, and experts were rushed in to explain that the lost tiles were in non-critical areas and to reassure the nation that the astronauts were in no danger.

After the privations of the Gemini and Apollo missions, Young could hardly fail to notice how much pleasanter life was aboard *Columbia*. Not only was the shuttle the first US spacecraft to use a 'normal' oxygen-nitrogen atmosphere at sea level pressure (14.7 psi); it was also equipped with a toilet, a galley and a wide variety of meals and drinks. The one drawback was that the men were expected to sleep in their ejection seats instead of sleeping bags, on call at all times in case some emergency arose.

After more than two days of travel around Earth, the crew struggled into their pressure suits and prepared for re-entry. Young turned the orbiter around so that it was flying tail first, then fired the manoeuvring engines to slow its velocity. *Columbia* hit the upper atmosphere east of Australia and began the 4400-mile-long glide towards California. Contact was lost for 16 minutes when the ship's exterior temperature rose to over 1500°C. As the American coast approached, Young took over to initiate a series of S-turns designed to slow the shuttle. There would be only one opportunity to reach the runway.

A double sonic boom startled the waiting crowds as *Columbia* swept across the dry lake bed at Edwards Air Force Base and turned for the final approach. The shuttle hit the 7-mile runway at 215 knots, the nose wheel came down seconds later, and *Columbia* trundled to a halt after 8993 feet. 'This is the world's greatest flying machine, I'll tell you that,' commented the normally laconic John Young.

## THE SPACE GARAGE AND LAUNCH PAD

Over the years the shuttle has been used as a launch pad for various commercial, scientific and classified military payloads (usually spy satellites or early-warning satellites). The first such launches took place during STS-5 (November 1982) and continued until the *Challenger* disaster.

The shuttle is not an ideal satellite launch vehicle, however. It flies at 200–250 miles above Earth, far too low for most satellites, and an additional strap-on motor is needed to boost a communications satellite to its operational orbit 22,300 miles above the equator. It doesn't make much commercial sense for NASA either. Even at $10 million or more a time, such commercial activity can barely dent the agency's $500 million shuttle

*The European Space Agency's EURECA satellite rests firmly in the grasp of the shuttle's remote arm as Atlantis passes over the Persian Gulf.*

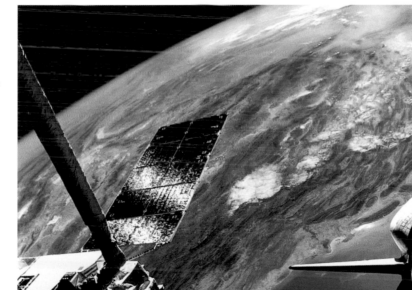

launch bill. On the other hand, it does take customers away from private companies. By the mid-1980s the US rocket industry was in a parlous state, largely because the shuttle monopolized the launch of federal government payloads.

Reliance on a manned launch system has also had an effect on some scientific missions. Before the loss of *Challenger*, the *Galileo* spacecraft was to be launched from the shuttle towards Jupiter by a powerful Centaur rocket stage. Following a safety review, this was considered too dangerous, and an alternative method had to be found. *Galileo* eventually got under way in October 1989.

The grounding of the shuttle for more than two years also revealed to NASA the folly of putting all its eggs in a single launch basket. Major projects were delayed for several years until a new launch opportunity could be provided. For example, the Hubble Space Telescope, originally scheduled for lift-off in August 1986, got off the ground in April 1990, while the Ulysses probe set off more than four years late.

If the shuttle has its limitations as a launch vehicle, it has certainly proved its worth on occasions as a space garage. With the aid of its robot arm, a number of satellites have been 'dropped' overboard for retrieval later. One of them, the Long Duration Exposure Facility, was exposed for much longer than originally anticipated. Lifted over the side by the robot arm in April 1984, it drifted around Earth, battered by radiation and space debris, for almost six years until it could be retrieved. Other satellites whose motors have failed to operate properly have also been brought into the shuttle's cargo bay, either for in-orbit repair or for a return to Earth where they can be checked and reused.

Aware of the shuttle's unique capabilities, designers have also begun to modify satellites so that they can easily be refurbished by spacewalking astronauts. The first of these was the Solar Max satellite, which had been largely inoperative after three fuses blew in its attitude control system in November 1980. Hauled aboard *Challenger* in April 1984, the 'Ace Satellite Repair Co.' of George Nelson and James van Hoften replaced the faulty control box and slotted in a new electronics system.

Perhaps the most famous example of in-orbit repair came in 1993 when the Hubble Space Telescope was brought aboard to correct its flawed mirror and shaky solar panels. Equipped with some 200 tools and other aids, a team of four astronauts shared the difficult, intensive work over five exhausting spacewalks, lasting $35\frac{1}{2}$ hours. Since then, a second, routine, servicing mission has further upgraded its performance.

## SPACELAB AND SPACEHAB

In the 1970s the European Space Research Organization (ESRO), the predecessor of the European Space Agency (ESA), expressed a desire to take part in the US manned spaceflight programme by providing a laboratory that would fly in the shuttle's cargo bay. In return, European scientists would gain access to a unique facility for experimentation in microgravity. Deprived of a space station by financial cutbacks, NASA was happy to accept the Spacelab as a substitute that would allow microgravity research and expand the shuttle's living space.

The manned version of Spacelab consisted of a pressurized laboratory inside which astronauts could work on experiments in a shirtsleeve environment. Depending on mission requirements, the 13-foot wide lab was available in two sizes. The short module was 14 feet long, while the long version was 23 feet long. One to three pallets on which experiments were exposed directly to the vacuum of space could be added.

Spacelab first flew in November 1983, with German Ulf Merbold, the first non-

*Robert Parker (left) and German Ulf Merbold make light work of their experiments during the first Spacelab mission in 1983. Biomedical sensors on their skin record their body condition. A loop on the 'floor' is used as a foot restraint when an astronaut needs to remain in one spot.*

American to fly on a US spacecraft, as the ESA representative. For the first time, astronauts worked through two 12-hour shifts as they tended 72 experiments.

Since then, various versions of Spacelab have flown on 22 occasions, including two missions sponsored by Germany and one mainly devoted to Japanese research. Science missions have covered many disciplines, from bio-medical investigations involving two monkeys and 24 rats, astronomy missions equipped with various telescopes, materials processing missions intended to produce new alloys and medicines, and life science missions to study how humans adapt to weightlessness. The experiments are usually mounted on racks in the walls, which are designed for easy removal and change over between flights.

Many of the experiments in Spacelab involved studies into the behaviour of fluids in zero gravity. Scientists were interested to discover how to create a uniform mixture of the liquids that form certain metal alloys such as those used in bearings or superconductors. Studies of crystal growth in on-board furnaces enabled metallurgists to make better metals, alloys and electronic materials.

Other experiments grew protein crystals for drug researchers. Many of these grow larger in space and have a more uniform structure than when grown on Earth. Information gained from X-ray analysis of the crystals may ultimately lead to the development of more effective drug treatments for many diseases. NASA claims that such research has already provided important advances in the understanding of many diseases including AIDS, heart disease, cancer, diabetes, anaemia, hepatitis and rheumatoid arthritis.

With the construction of the ISS, Spacelab is no longer needed. The last Spacelab mission devoted to materials processing was the Microgravity Science Laboratory flight (STS-94) in July 1997. Its final bow came in April 1998 with the STS-90 Neurolab mission, which was devoted to studying the effects of zero G on the human nervous system. Included in its crew were three medical doctors and an assortment of fish, snails, crickets, mice and rats.

Spacelab was a large, heavy facility, which filled much of the cargo bay. A scaled-down laboratory is now used instead for orbital research on certain occasions. The first Spacehab flew in June 1993, carrying 21 materials and life sciences experiments. Despite a lack of commercial customers other than NASA, Spacehab has now become a familiar feature on recent shuttle flights. As with Spacelab, it may be launched in single or in double format. The 10-feet long, single module doubles the workspace provided by the shuttle.

Earth observation is also an important part of shuttle research. One of the most innovative remote sensing experiments was the use of a radar system to see through the clouds and probe surface and subsurface features. Major successes included the discovery of ancient river channels beneath the Sahara Desert and of ancient Arabia's 'lost' city of Ubar.

## SHUTTLE SCARES

Ever since *Columbia* lost some protective heat tiles from its outer skin during launch in 1981, the shuttle programme has had to endure more than its fair share of safety scares and technical problems. Despite the post-*Challenger* safety review, further issues of reliability continue to surface.

In 1990 the entire fleet was grounded for 5 months while engineers struggled to correct hydrogen leaks in both *Columbia* and *Atlantis*. 1991 was also a bad year. STS-39 was postponed for 7 weeks when hairline cracks were found in the hinges of the propellant line doors of all four orbiters, including *Endeavour*, which was still under construction. Later that year, further delays occurred after a number of cracked temperature sensors were discovered. If fragments from these had been sucked into the SSMEs, the result could have been disastrous.

More recently a launch abort was barely avoided during the February 1996 launch of STS-75, when one engine's pressure gauge read only 45 per cent. The crew were preparing to attempt a dangerous return to Florida when ground controllers confirmed that the engines were running properly.

In March 1996 the orbiter *Atlantis* sprang a leak in the hydraulic system that supplies power to its wing flaps, rudder and landing gear. Mission controllers considered bringing the craft home early. Then sensors wrongly reported that the two cargo bay doors had jammed shut. If correct, this could have caused overheating and severe damage to the shuttle's electronics. The orbiter eventually limped home with several thrusters non-operational.

Sceptics argue that such incidents are still far too common for the shuttle ever to be regarded in the same light as an operational airliner. Although there have been no craft lost or crew fatalities in the last 10 years of operations, the question of safety is far from academic. A recent risk analysis commissioned by NASA stated that, despite continual modifications to the fleet, there is a 50 per cent chance of another catastrophe before the year 2015. The report concluded that the most likely scenario is a launch failure caused by the explosion of an SSME. Another possibility is a crash on landing caused by failure of the landing gear and brakes. Certainly, there is a fatalistic air about those involved in the programme.

'There is going to be another accident,' said former Air Force Secretary Peter Aldridge. Former astronaut and NASA administrator Richard Truly agreed. 'Sooner or later, something's going to happen. It doesn't have to be a *Challenger* accident. It could be a nose wheel failure.'

Could NASA survive another shuttle disaster? On the previous occasion, the fleet was grounded for more than two years and Congress agreed to fund a replacement vehicle at a cost of $2 billion to the US taxpayer.

'To me, it's not a political issue,' said NASA chief, Dan Goldin. 'It is a statement about the will of the American people.'

## RISING COSTS AND FALLING EXPECTATIONS

The record for shuttle flights still stands at nine in 1985. Even this flight rate put the agency under such pressure that it led to the loss of *Challenger* the following January.

*Russian Elena Kondakova uses the Spacehab glovebox during mission STS-84 in May 1997. Kondakova was only the third Russian woman to fly in space. In the background, mission specialist Carlos Noriega checks an experiment.*

Under current plans, the schedule is expected to peak again at nine by the year 2000 during construction of the ISS.

Meanwhile, despite a programme to improve efficiency that saw annual operating costs fall by about $1 billion between 1992 and 1995, NASA still spends about $3 billion a year on the shuttle programme, approximately one-fifth of its entire budget. Some 17 years after the birth of the shuttle age, the agency has been forced to come to terms with an ageing fleet, a declining workforce and budget, and falling morale.

Faced with the unpalatable facts that each mission costs about $400 million and no replacement vehicle will be available until well into the next century, Dan Goldin came to the controversial conclusion that the way ahead was to privatize the shuttle programme. Thousands of workers lost their jobs as responsibility for shuttle operations was transferred to a newly formed private company called United Space Alliance, a joint venture by US space giants Lockheed Martin and Boeing.

Goldin remained unrepentant, dismissing the idea of a private corporation risking lives to increase profits. 'With Lockheed Martin and Rockwell [now part of Boeing] we have two experienced companies that clearly understand how to operate the shuttle safely,' he said.

## INTO THE FUTURE

The shuttle fleet has gradually been modernized and updated since *Columbia*'s debut in 1981. In particular, when *Challenger* was replaced by *Endeavour*, NASA took the opportunity to update its fifth operational orbiter with the latest hardware, including more powerful computers, improved navigation systems and auxiliary power units, plumbing and electrical connections to permit stays in orbit of up to 28 days, and a drag chute to slow the craft on landing. Most of these modifications have since been introduced on the other vehicles.

When the shuttles were built, they were intended to fly at least 100 missions each. Since the existing orbiters have recorded barely a quarter of that number, they could, in theory, continue to fly until the year 2030. However, if the shuttle is to remain a useful, reliable machine without constantly draining NASA's dwindling financial resources, it must be continually upgraded and maintained.

A number of hardware and software improvements have been proposed or are already in the pipeline. Under pressure to reduce lift-off weight in order to make room for extra payload, NASA's Marshall Space Flight Center at Huntsville, Alabama, has constructed a new, super-lightweight, external fuel tank for the shuttle. The new model is the same size as the one currently used but is about 7500 pounds lighter thanks to its aluminium lithium construction and newly designed walls.

Starting with *Atlantis*, the fleet will also undergo a redesign of the main display panels. Flat panel displays will replace the 1970s-era cathode ray tube screens and switches, making it much easier to monitor spacecraft systems and fly the shuttle.

US space policy has also altered direction. After the *Challenger* disaster, the shuttle was taken off mundane duties, such as launching US DOD payloads and commercial satellites. However, priorities and opinions change. NASA and United Space Alliance want to open up the shuttle to new, paying customers. They envisage the orbiters flying 9–12 times a year, although to achieve this, orbiter processing time would have to fall from 80 to 50 days. The US Air Force is also considering the feasibility of once again placing payloads on shuttles.

An inevitable outcome would be a modified maintenance schedule. At present orbiters are checked and upgraded after eight flights or three years, whichever comes first. No more than one vehicle can be handled at a time. Such guidelines would have to be relaxed if the flight manifest were to become more aggressive.

## BURAN: THE SOVIET SNOWSTORM

During the Cold War, if one superpower gained what was seen as a technological advantage, the other superpower felt obliged to respond. Soviet military authorities saw the US shuttle being used for top secret missions and regarded it as a potential threat. The obvious (but extremely expensive and difficult) answer was to build a Soviet version.

Throughout the 1980s Western intelligence reports told of a Soviet spaceplane programme, and photographs obtained by the Australian Air Force showed a small, delta-winged craft being recovered from the Indian Ocean.

The *Buran* shuttle, which was revealed to the world on 29 October 1988, bore a remarkably close resemblance to its American counterpart – too close some said. This was hardly surprising since little of the NASA shuttle technology had been kept under wraps and it was available for all to see. The most significant difference was the launch system. Rather than depend on SRBs, which had been the cause of the *Challenger* disaster, the Soviets decided to launch *Buran* attached to the side of a giant rocket called Energia. The two would separate at an altitude of 70 miles, after which the shuttle would use its own manoeuvring engines to achieve a 155-mile-high orbit.

Although the 29 October countdown had to be terminated less than a minute before launch when a support arm separated too slowly from the body of the booster, the second attempt on 15 November went like clockwork. On only its second test flight, Energia successfully carried the delta-winged craft into a low Earth orbit. Then, 205 minutes and two circuits of Earth later, the unmanned shuttle made a perfect automatic landing on the runway at Baikonur Cosmodrome, just 7½ miles from its starting point. Although its white exterior was singed by the heat of re-entry and four of its ceramic heat tiles had been dislodged, a new era in Soviet space exploration seemed to beckon.

Nothing could have been further from the truth. Energia and *Buran* never flew again. Having spent billions of roubles on development, no one could think of anything to do with them that was useful or cost-effective. Plans to dock *Buran* with the *Mir* station and carry out a crew transfer never materialized. After years of procrastination, the government decided that the projects were too expensive and pulled the plug. Both *Buran* and a second orbiter, which was intended to have been equipped with ejector seats and a life-support system for crewed flight, were mothballed.

Although *Buran* was almost identical in size and shape to its American counterpart, to dismiss it as a space shuttle clone is to ignore the long history of spaceplane studies in the Soviet Union. In fact, the Soviets had begun preliminary work on such a vehicle as early as the mid-1960s, although work on *Buran* had not started until 1974. Work was sufficiently advanced by 1978 for recruitment of the first group of test pilots. A series of suborbital trials, then four orbital tests, were carried out on a sub-scale model, beginning in 1982 with *Cosmos 1374*. Finally, as in the American shuttle programme, between 1985 and 1988 cosmonauts took part in approach and landing trials using a full-scale mock-up equipped with jet engines.

Only two of the *Buran* cosmonauts ever flew in space, the pair who had been selected as the prime crew for the craft's first manned flight. Igor Volk spent a week aboard the

*Salyut 7* space station in 1984, while Anatoli Levchenko spent a similar time aboard *Mir* in 1987. On his return to Earth in a Soyuz capsule, Levchenko was immediately transported by helicopter to a nearby airfield. Within half an hour, he was flying a TU-154 airliner to Moscow while physicians tested his reflexes.

## THE X PROGRAMMES

Since the early days of rocketry, engineers have dreamed of building a completely reusable spaceplane that will take off and land from a runway. During the 1960s significant progress was made in the US by experimental lifting body craft and air-launched rocket planes, such as the X-1 and the X-15.

Today, as the shuttle fleet begins to show its age, NASA and aerospace companies are ploughing millions of dollars into the development of space transportation systems for the twenty-first century. Inspired by the promise of the X-prize, a $10 million reward offered for the first privately funded team to launch a craft capable of carrying three adults to the edge of space, 62 miles above Earth, several companies are challenging NASA's monopoly of reusable space technology.

Lifting people to this height for a brief glimpse of a curved blue horizon is one thing. Staying aloft long enough to perform useful work is another. Designs for new vehicles vary greatly, but few anticipate carrying passengers or crew until well into their evolutionary development.

One, known as Roton, is described as a 'space-helicopter', since it would be equipped with a rocket motor on the end of each of its four rotor blades. Another, the Eclipse spaceplane, would be towed by a Boeing 747 aircraft to an altitude of 9 miles where it would fire its rocket engine to attain orbit. A third, known as Pioneer, would use conventional jet engines to take off, then take on liquid oxygen rocket fuel from a tanker aircraft before completing its journey.

On the government side, the cornerstones of NASA's reusable launch vehicle (RLV) programme are the experimental X-33 and X-34. Eventually, the agency hopes to slash the cost of putting payloads into space from $10,000 a pound to $1000 a pound.

Plans call for the unmanned X-34 to fly up to 25 times within a year. The suborbital vehicle will be launched from an L-1011 airliner, reaching speeds of 6000 mph and flying at altitudes of approximately 50 miles. At mission's end, it should be able to touch down on a runway through rain, fog or cross-winds of up to 20 knots.

However, NASA's flagship vehicle is the X-33, a small prototype of a single-stage-to-orbit RLV. Using a new form of rocket propulsion known as linear aerospike engines, the wedge-shaped X-33 is scheduled to begin test flights in 2000. Launched vertically from Edwards Air Force Base, it will fly at altitudes of 60 miles and speeds exceeding 11,000 mph before landing on runways in Utah and Montana.

X-33 demonstration flights and ground research should provide sufficient information for private industry to decide by the year 2001 whether to proceed with production of a full-scale, commercially developed RLV. If all goes well, Lockheed Martin hopes to develop the *VentureStar* early in the next century. Seven aerospike engines will eventually be used to power the full-scale vehicle into orbit. Although primarily foreseen as a cheap method of delivering supplies to the ISS and launching satellites, the addition of a crew transfer module may enable *VentureStar* to evolve into a replacement for the shuttles during the first decade of the next century.

# 6

# WALKING IN A VACUUM

## THE ULTIMATE VIEWPOINT

'I love EVA. I think it's a beautiful experience to be out there in a suit with just a face plate between you and the vacuum. The Earth is an even richer colour than you can imagine. It's a fantastic view.' So said astronaut Michael Foale. The irony is that, despite the enormity of the experience, most spacewalkers complain about the lack of opportunity to pause and stare. With a full schedule of experiments to deploy or satellites to repair, the moment passes without the opportunity to appreciate it fully.

On the following page is cosmonaut Valentin Lebedev's description of his first venture into open space on 30 July 1982. Although dozens of astronauts and cosmonauts have repeated Lebedev's experience, it remains one of the most evocative accounts of this unworldly adventure.

*'A beautiful experience', but Carl Meade and Mark Lee barely have time to admire the view while they test the SAFER rescue system 150 miles above Earth.*

*Once the exit hatch was opened, I turned the lock handle and bright rays of sunlight burst through it. I opened the hatch and dust from the station flew in like little sparklets, looking like tiny snowflakes on a frosty day. Space, like a giant vacuum cleaner, began to suck everything out. Flying out together with the dust were some little washers and nuts that had got stuck somewhere; a pencil flew by.*

*My first impression when I opened the hatch was of a huge Earth and of the sense of unreality concerning everything that was going on. Space is very beautiful. There was the dark velvet of the sky, the blue halo of the Earth and fast-moving lakes, rivers, fields and cloud clusters. It was dead silence all around, nothing whatever to indicate the velocity of the flight ... no wind whistling in your ears, no pressure on you. The panorama was very serene and majestic.*

## ABOVE THE ATMOSPHERE

Look closely at a photograph of Earth taken from space. Barely visible above the planet's limb is a thin blue line – the atmosphere. Compared to the vast bulk of our world, the life-sustaining gases that we breathe are squashed into a wafer-thin skin about 75 miles deep. Above this is the vacuum of space.

The presence of this blanket of gases is usually taken for granted. We tend to forget that, without the oxygen that makes up 20 per cent of this blanket, we could not survive. We also forget that air gets thinner with altitude. At altitudes above 63,000 feet (12 miles) the air is so thin and the amount of oxygen so small that humans cannot survive without an oxygen mask and a suit that maintains pressure around the body.

Inside spacecraft space travellers can survive quite happily without pressure suits. In the background, however, there is always the lingering fear of an air leak. The designers of the first spacecraft had no illusions about the dangers of depressurization, and, with the exception of the Russian cosmonauts who flew in the first Voskhod and Soyuz craft, all the early crews wore pressure suits. Although they were abandoned after the first shuttle flights, the suits were reintroduced in 1988 when flights resumed after the *Challenger* disaster. Ever since, shuttle crews have taken off and landed shrouded in the safety of their orange space gear.

## ALL DRESSED-UP WITH NOWHERE TO GO

The first spacesuits were based on garments worn by military pilots and were used solely as insurance against sudden depressurization of the cabin. By the time the first astronauts were ready to venture above Earth's gaseous envelope, reconnaissance aircraft such as the U-2 and X-series rocket planes regularly flew to the edge of space.

During the suborbital and orbital Mercury missions, American astronauts wore adaptations of the US Navy's MK-IV pressure suit, manufactured by B.F. Goodrich of Akron, Ohio. These suits were little more than inflated, man-shaped rubber balloons beneath a layer of fabric and insulating material, although the astronauts had to remember to fasten 13 zippers and three rings before they were snugly cocooned inside. Each suit was tailor-made for its occupant. To ensure a perfect fit, each individual had to strip down to his long johns at the factory and be covered in strips of wet paper. When it dried, the paper created a mould for the final suit. Complete with helmet, the suits weighed 22 pounds and cost $5000 each.

All pressure suits are made up of a number of different layers. The suit worn by the Mercury crews consisted of an inner layer of Neoprene-coated fabric covered by a restraint layer of shiny aluminized nylon. It was usually worn 'soft' or unpressurized so that movement was somewhat easier. However, it was a different proposition when fully pressurized.

Joint mobility at the elbows and knees was provided by fabric rings sewn into the suit. When the joint was bent, the rings folded in on themselves, reducing the suit's internal volume and so increasing pressure. Both straight- and curved-fingered gloves proved a handicap when pressurized, so the designers came up with a compromise: the middle finger on the left hand was made straight, while the remainder were curved.

Limited mobility was not a serious handicap in the tiny Mercury capsule, but the crew took the precaution of taking along a small rod with a hook on the end for pulling at levers and a stub for pushing at buttons. The astronauts called it a 'swizzle stick'.

Oxygen circulating through the suit helped to maintain a comfortable body temperature. It could be adjusted anywhere between 10 and 32°C – Alan Shepard preferred it cool (about 15°C), while others liked a sub-tropical 21°C. The gas entered through an inlet valve near the waist and flowed through the underwear before leaving via an outlet near the right ear in the helmet. As it left the suit, it carried with it any body odours, perspiration, nasal discharge, bits of hair and waste gases. The 'tired' gas was then filtered, purified and cooled before flowing back into the suit again.

Oxygen for both the suit and the cabin came from the same source. If the cabin sprang a serious leak, the supply would be routed entirely through the suit. If the suit failed, the astronaut could open the helmet faceplate and breathe the air in the spacecraft. If both events occurred simultaneously, the astronaut could switch to an emergency back-up supply of oxygen that would last for 80 minutes, enough time to make an emergency splashdown.

The exterior of the Mercury helmet was made of hard fibreglass, while the interior consisted of an outer covering of leather, a layer of soft 'comfort foam' and a thicker layer of stiff 'crash foam'.

## OFF-THE-PEG RUSSIAN SUITS

Spacesuits had a similar background in the Soviet Union, where military hardware was adapted for the first cosmonauts. The Vostok cosmonauts wore a more bulky and slightly heavier (22–26 pound) garment than their American counterparts.

Gagarin and his comrades had a blue woollen comfort layer and a constant-wear biomedical garment next to their skin. On top of this was the three-layered pressure suit. First came a zippered, sky blue heat insulation layer, complete with a ventilation system of tiny air pipes. Next was a thin, blue rubber layer, which acted as a gas 'bag'. The uppermost layer of the pressure suit was made of Dacron, and it employed a series of hinges and ropes to give an accurate fit for each occupant. All this was hidden beneath a zippered coverall, which was coloured orange to ease recognition by rescue teams after an emergency landing.

Black lace-up boots were worn to protect the ankles and feet. Made-to-measure gloves locked onto the sleeves of the suit, and a hard white helmet locked onto the upper torso. The visor could be raised or lowered according to different lighting conditions. A feed port was provided in the lower part of the helmet so that the cosmonaut could drink with the visor in place. The visor would automatically lower if there was a drop in cabin pressure.

Today's Sokol suits, although less bulky, are still quite uncomfortable. Like the suits worn by shuttle crews, they are worn only for lift-off and re-entry. British cosmonaut

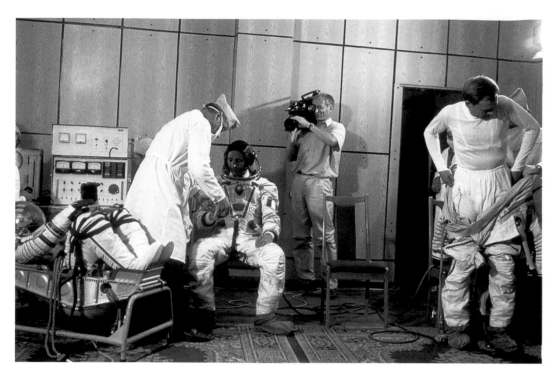

Pre-flight suiting up for the crew of Soyuz TM-15. Anatoli Solovyov, Sergei Avdeyev and Frenchman Michel Tognini appreciate all the help they can get in donning their Sokol pressure suits.

Helen Sharman found the suit to be 'one of the most unwieldy and inconvenient garments imaginable'. Certainly, it is not ideal for walking around in normal gravity, partly because a bulky ventilator has to be carried by the occupant to circulate air through the suit and partly because the internal wire reinforcement makes motion difficult. The cosmonauts keep the helmet open to fresh air or cabin air for as long as possible. Sharman commented: 'The only time you feel at all comfortable inside it is when you are in the position for which it was designed to be used: lying in the capsule on your back with your knees up.'

During pre-launch pressure tests on Earth, the Sokol is pumped up to 1.44 atmospheres, but at this level the cosmonaut is barely able to move. There are two normal settings for use in space: 0.44 atmospheres, which is safer but more rigid, and 0.27 atmospheres, which offers less protection to the wearer but allows rather more freedom of movement.

## WALKING IN SPACE

The hostile nature of outer space obliges human space travellers to seek protection at all times or perish. For most of their time aloft they have to be content to live inside the spacecraft, occasionally glancing at Earth and stars through a nearby window.

Spacewalking is different. Encased in a pressure suit, the human body is free to float 200 miles above Earth, suspended between the blackness of space and the blue and white planet. Sweeping around the planet at 17,500 mph, the spacewalker travels effortlessly from day to night and back again, witnessing a sunrise and sunset once every 90 minutes. It is the nearest a human being can come to being a moon or a meteorite – a free-floating child of the universe. Whether the spacewalk – or extra-vehicular activity (EVA) as it is officially known – lasts for an hour or for 6 hours, it will stay in the memory for a lifetime.

As Leonov discovered, it was one thing to provide temporary protection against emergency depressurization of a spacecraft, but designing a suit for EVA was a different

## The First Walk in Space

The first spacewalk was an ambitious attempt by the Soviets to beat the Americans, but it almost ended in tragedy. A member of the first cosmonaut group, 30-year-old air force officer Alexei Leonov, was the man chosen to make history. For nearly a year Leonov acquainted himself with the Voskhod capsule and its unique telescopic airlock, a cylinder 6 feet long and 3 feet in diameter. Training included taking part in vacuum chamber tests and practice 'exits' from a mock-up craft during brief moments of weightlessness as they flew parabolic curves in Tu-104 aircraft.

On the night before the launch, Korolev paid a visit to Leonov and his crew commander Pavel Belyayev. 'If you spot some trouble – anything might happen – don't take unnecessary risks.' Korolev's words were more prophetic than he realized.

*Voskhod 2* lifted off from Baikonur on the morning of 18 March 1965. Once in orbit, the crew prepared for the giant leap. By breathing pure oxygen at 6 psi instead of the normal oxygen-nitrogen mixture at 16 psi, Leonov was able to purge all the nitrogen from his blood and respiratory system. Towards the end of the first orbit, the cosmonauts made a final check of their suits and the now extended airlock. Leonov then floated from his couch into the airlock, and Belyayev closed the door behind him. Safely inside, Leonov waited patiently until all the air in the airlock was vented into space. It was time to open the outer hatch and stare in amazement at planet Earth 300 miles below. Leonov later described the view:

*The stars were bright and unblinking. I could distinguish clearly the Black Sea with its very black water and the Caucasian coastline ... . I saw the mountains with their snow tops looking through the cloud blanket covering the Caucasian range.*

After installing a cine-camera to record the moment for posterity, Leonov inched his way out of the airlock into the vacuum of space. 'I'm pushing off,' he cried as flickering black and white pictures back on Earth showed the white-suited cosmonaut launch himself head first from the tunnel at the end of a 16-feet tether. Inside this lifeline were an emergency oxygen feed, a telephone cable and wires carrying telemetry from sensors attached to his body.

Belyayev gave him cautious encouragement. 'Don't be in a hurry, Lyosha. Do it as you were taught.' Arms outstretched, he drifted in open space, restricted only by the length of his tether. After 10 minutes of cavorting in space, the call came to bring the somersaults to a close. Leonov reluctantly complied. But ecstasy turned to concern as things began to go wrong.

Hampered by his stiff, bulky gloves, Leonov had to struggle to remove the cine-camera from its bracket. Then it tried to float out of the airlock as the cosmonaut struggled to manoeuvre his feet through the entrance. As he fought against the rebellious camera, the confined room and a ballooning space-suit, there was a serious possibility that he might be unable to reach the Voskhod cabin. Despite the risk of contracting the 'bends', a desperate Leonov steadily reduced the air pressure in his suit to 0.25 atmospheres. To his relief, this deflated the suit and improved its flexibility. Nevertheless, by the time the cosmonaut managed to squeeze head first inside, his pulse had soared to 168 and he was perspiring profusely.

Leonov's official EVA time was 23 minutes 41 seconds, of which about 12 minutes were taken up with the spacewalk itself. The inflatable airlock was jettisoned and the duo prepared for re-entry.

*Alexei Leonov becomes the first human to leave the safety of his capsule and walk in space.*

proposition. Should any part of the suit fail during a spacewalk, there was no back-up system. A dead astronaut would simply have been cut free and left to drift around Earth for many years before his body was cremated during re-entry.

Not only must an EVA suit be able to shield against the vacuum of space, temperature extremes and meteorite impacts, it must also be durable and allow freedom of movement. If the final frontier was to be conquered, mankind would have to learn how to work and manipulate tools or large objects. Unfortunately, the two requirements worked against each other. Turning the spacewalker into a modern equivalent of a knight in armour was not conducive to improving dexterity and flexibility.

Designers of the first American EVA suits used in project Gemini overcame the problem of protection by adding extra layers. Moving out from the skin, the astronauts wore: long cotton underwear; a nylon comfort liner; a pressure bladder of Neoprene-coated nylon; a restraint layer made of a fishnet-like, one-way stretch fabric called Link Net; a layer of felt to help keep the wearer warm at night and give added protection from meteorite strikes; seven layers of aluminized mylar insulation; and a cover of white, high temperature nylon. The Link Net was added to enhance the mobility of the Gemini suits, particularly when they were unpressurized, while the layers of insulation were to protect against temperatures ranging from +120°C at midday to -120°C at night.

## WHITE'S WALK

After Leonov's pioneering effort, NASA engineers were under pressure to bring forward their own plans for a man to exit the Gemini spacecraft. The first US spacewalk was approved on 24 May 1965, just 10 days before the scheduled launch date, when the new hardware passed its qualification tests.

Once in orbit, *Gemini 4* astronauts James McDivitt and Edward White were determined not to rush their preparations and postponed the EVA for one orbit while they carefully went through the checklist. Eventually, the crew lowered cabin pressure to 2 psi in order to check their suits. Assured that all was well, McDivitt vented the remaining cabin air, and White opened his hatch as they passed over Hawaii. After checking three times that he had removed the lens cap from the movie camera, he was ready for his leap into the void.

Aided by a small hand-held thruster pack, White made a controlled turn to the left and moved alongside the spacecraft by squirting small spurts of compressed oxygen. When the manoeuvring unit ran out of gas, he released the remainder of his 25-feet tether from its stowage bag and adjusted his motion by tugging away on the snaking line.

The gold-coated umbilical cord acted as both a lifeline and a communication link. Inside the 1-inch diameter cord were a strong nylon tether, an oxygen hose, four electric leads and a communications lead. If any problem arose with the oxygen link to the spacecraft, White would be able to draw on an emergency supply from a small pack worn on his chest. Fortunately, this was not necessary, and it was a disappointed astronaut who reluctantly obeyed the call to climb back in after 21 minutes.

Ironically, as with Leonov, the most difficult part of the EVA was its final act. After gingerly handing the umbilical and manoeuvring unit to McDivitt, White was unable to close the hatch. There followed a tug of war, during which White's pulse raced to 178, and by the time the 15-minute struggle was concluded, both men were bathed in perspiration.

Opposite *Edward White cavorts for the camera during the first American spacewalk. In his right hand is a small thruster unit to help him manoeuvre. The 25-feet gold-taped tether both carried his oxygen supply and acted as a safety line. The pack on his chest carried an emergency oxygen supply.*

## PREPARING THE ROAD TO THE MOON

Both the Soviets and the Americans knew that the Moon Race could not be won without conquering the hidden devils of EVA. While NASA was able to build on White's pioneering effort with a series of ambitious spacewalks during later Gemini missions, Leonov's difficulties and the limited capabilities of the Voskhod spacecraft led to lengthy delays in the Soviet programme.

Not that all the Gemini EVAs were spectacular successes. America's second walk in space had to be terminated early when Eugene Cernan's suit began to overheat as he struggled to put on a huge backpack called an astronaut manoeuvring unit. As the sweat poured from his body and his pulse raced to 180 beats a minute, the suit's environmental

control system was unable to cope. Confronted with a fogged visor and garbled communications, Cernan was recalled to the cabin, where he spent another 15 minutes wrestling with the 25-foot long umbilical before he could close the hatch.

The *Gemini 10* and *11* spacewalks involved activities with an Agena rocket upper stage. Michael Collins was given the unenviable task of launching himself across a 10-foot gap to grasp the Agena's slippery metallic skin. With the aid of a hand 'gun' and a 50-foot tether, Collins was able to succeed at the second attempt. Unfortunately, the photographic record of his exploits was lost when his Hasselblad camera floated off into the void.

On *Gemini 11* Dick Gordon was asked to attach a tether to the docked Agena. Once again, the lack of hand- or footholds added to the difficulty of the task, and although Gordon succeeded in sitting astride the booster on his 'bronco ride', he was ordered back inside the spacecraft when his suit began to overheat.

NASA officials learned a great deal from these setbacks. On the final Gemini mission, Aldrin was able to take advantage of a telescopic handrail and improved tethers and restraints. During three successful EVAs, the future Moonwalker set a new overall record of 5 hours 32 minutes outside his spacecraft.

The final preparations for the Moon landing were completed on the *Apollo 9* Earth-orbiting mission, when Russell Schweickart and David Scott tried out the lunar pressure suit and its portable backpack.

Meanwhile, the Soviets were struggling. Since there was no docking tunnel on the new Soyuz vehicle, a cosmonaut on a Moon mission would have to exit his spacecraft and transfer through open space to and from the Moon lander. Plagued by the *Soyuz 1* disaster and a series of docking failures, the only occasion this procedure was actually attempted took place in 1969 when two cosmonauts successfully walked from one Soyuz to another. No further Soviet spacewalks took place until 20 December 1977.

## DESIGNING A SPACESUIT

Suit technology has improved over time, but modern design still involves a trade-off between increased complexity, cost, weight and reduced flexibility. Michael Collins commented: 'You had to work hard to bend a pressurized Gemini suit. The Apollo suit was more mobile but not as comfortable. The early ones, especially, dug into shoulders or crotch, or both.'

Keeping the body cool and allowing the skin to breathe are far from easy encased in an airtight suit. EVA suits must be airtight yet allow the wearer to perspire and breathe. The first air-cooled suits had not been too successful, so for Apollo and subsequent programmes the cooling problem was solved by circulating water through a network of tiny tubes sewn inside a pair of long johns. Oxygen tubes were also added to this garment. In this way, circulating oxygen could help cool the astronaut, so reducing the likelihood of fogging on the helmet visor and carrying away exhaled carbon dioxide.

Nowadays, astronauts come in all shapes and sizes. NASA has compromised by making a limited number of off-the-peg sizes for the shuttle EVA crews. There are five standard sizes of torso, a variety of sleeve and trouser lengths and a single-size helmet. Only the gloves are custom-made, and they cost $20,000 each. Each piece of clothing can be separated, cleaned, dried and re-used by other crews.

Climbing into the two-piece suit is fairly straightforward and can be done unassisted in about 5 minutes, compared with an hour for the Apollo crews. Once the legs are inserted into the trousers, the astronaut simply raises his hands above his head and pushes upwards into the upper torso. The suit is sealed by metal rings, which lock mechanically at the waist, helmet and gloves.

Size is not usually a problem for the Russians because the Orlan suit can be adjusted by means of straps. The one, unlucky exception was 5-foot-3-inch astronaut Wendy Lawrence, whose trip to the *Mir* space station was cancelled only weeks before lift-off because she was too small to fit the suit.

The current version of the Russian EVA suit, the Orlan-M, is the latest in a long line of suits developed by the Zvezda company. Three Americans have tried out the Russian EVA suit in orbit. After its first outing in space in April 1997, cosmonaut Vasili Tsibliev and astronaut Jerry Linenger praised its flexibility and helmet visibility compared with previous types.

The basics have remained the same for all Russian EVA suits since the 1970s. While the suit remains unpressurized, the occupant feels squashed in like a sardine with his head up against the helmet. Comfort levels improve once the suit is pressurized when oxygen, blown by a fan, circulates around inside. This also helps cut down condensation, particularly on the visor, and keeps the occupant cool. The Orlan suit is equipped with two oxygen tanks, a prime and a back-up for use in emergencies. Exhaled carbon dioxide is removed by a lithium hydroxide scrubber, and any moisture is condensed and removed.

Further body cooling is carried out by water circulating in 250 feet of plastic tubing sewn into an undergarment. Inlet and outlet fittings from these tubes are attached to the main suit, which has a pump to circulate the fluid.

Other in-built systems reinforce the comparison with an independent spacecraft. The suit has its own battery power supply and a radio for communications, and telemetry about pulse, breathing rate and body temperature is radioed back to mission control. Suit pressure gauges, control panels and a caution/warning panel are located on the upper chest area for easy viewing.

## LANDMARKS IN SPACEWALKING

Modern suits allow their occupants to operate outside the spacecraft for 8–9 hours, but it is rare for mission planners to schedule such an extended period of extra-vehicular activity. The longest Russian spacewalk was carried out by *Mir* cosmonauts Anatoli Solovyov and Alexander Balandin on 17 July 1990. They were obliged to spend 7 hours 16 minutes outside the station in an unscheduled EVA to secure three damaged thermal blankets on the exterior of their Soyuz TM-9 ferry craft. The repairs to the insulation proved to be the least of their worries.

In their haste to start work, the crew released the safety latch on the *Kvant 2* exit hatch before the module had fully depressurized. As the remaining air burst through the opening, it slammed the hatch back against its hinges. When the weary crew returned to the hatch after completing most of the repairs, they found, to their alarm, that the door would not close. Only after a lengthy struggle did they manage to shut the hatch by applying brute force. The hatch remained out of commission for almost 6 months until a later crew attached a new hinge.

Most spacewalks have involved one or two participants. The exception to this was the mission to rescue an ailing Intelsat VI communications satellite in 1992. After the shuttle's rendezvous with the satellite, Pierre Thuot perched on the end of the robot arm and tried to attach a specially designed grappling bar. After three failures to clamp onto it, the satellite was wobbling more than ever.

A similar scenario was played out the next day. Drastic measures were called for. On this occasion, three astronauts squeezed into *Endeavour's* airlock and floated into their positions. As commander Daniel Brandenstein inched the shuttle closer and closer, the men prepared to make a concerted grab at their quarry. Finally, they seized the rotating giant and attached the docking bar. The hard part over, the crew manoeuvred the satellite onto its new rocket motor and bolted them together. The 100th spacewalk ended in triumph after a record-breaking 8 hours and 29 minutes. The following day, their hard efforts were rewarded as the Intelsat set off towards its proper operational orbit, 22,300 miles above the equator.

## THE ROBOT ARM AND THE SPACE CRANE

One of the perils of stepping into space is the possibility that it may be one giant leap too far. Any space traveller launching himself into the void is likely to keep going and never return to the module unless he is attached by a tether or equipped with some type of propulsive device.

Tethers and other forms of restraint have an additional value in zero gravity. As Sir Isaac Newton pointed out 300 years ago, for every action there is an equal and opposite reaction. Although objects in orbit may have no weight, they certainly retain their mass and momentum. While we give little thought to turning a screw or bolt in normal gravity, Newton's law takes on a new significance for would-be space mechanics. Early astronauts found themselves rotating as they struggled to turn errant pieces of

*Vasili Tsibliev works on the Strela boom during a spacewalk on 29 April 1997. Also visible on the* Kvant 1 *module is the Sofora truss, which carries a thruster unit to help control* Mir*'s attitude.*

metal. Mission planners soon learned that astronauts need all the footholds and handholds they can find.

Today, spacewalking astronauts are given a vast range of special hand tools and crew aids. They include handrails, handholds, transfer equipment, protective covers, tethering devices, grapple fixtures, foot restraint sockets, and stowage and parking fixtures. In one extreme example, the crew of STS-82, who were servicing the Hubble Space Telescope, carried more than 150 tools and special aids into orbit. This space support equipment hardware ranged from a simple bag for carrying some of the smaller tools to sophisticated, battery-operated power tools.

Much of the work would not be possible without some way of providing extra mobility. On the shuttle, this is offered by the Canadian-built robot arm (officially known as the remote manipulator system). Controlled by an astronaut on the shuttle flight deck, this 50-foot long mechanical arm has three joints, which allow a spacewalker to be lifted into just the right position. With feet wedged into restraints and a work platform by their side, astronauts have conducted major repairs on a number of satellites.

The Russian equivalent on the *Mir* space station is the Strela mobile boom, which can be operated by the spacewalkers. This 46-foot long telescopic crane was attached to the base block during a spacewalk in 1991 as an aid to moving solar panels from the *Kristall* module to *Kvant 1*. Instead of carrying bulky items around by hand, cosmonauts can attach them to the boom. They also use it to move each other from one module to another. While one spacewalker sits on the end of the pole, the other rotates two handles to swing it to the desired location.

In a typical operation, a cosmonaut exits from the *Kvant 2* airlock and slides to the base of the crane down the pole, which has been left in position nearby. Once his colleague has tied himself and his equipment to the end of the boom, the crane operator cranks the handles to swing it around to the intended work location. Finally, he floats back along the boom to join his partner.

## THE SPACE CHAIRS

The manned manoeuvring unit (MMU) was intended to enable an astronaut to work untethered more than 100 feet from the spacecraft. It weighed 300 pounds on Earth and carried enough oxygen and power to supply its occupant for 7 hours. By careful use of its 24 tiny jets, which propelled nitrogen gas, the MMU could move up, down, forwards, backwards and even trace circles in the sky. Top speed was around 66 feet a second.

Its first in-flight trial came on 7 February 1984 when Bruce McCandless demonstrated his expertise in *Challenger*'s cargo bay. 'That may have been one small step for Neil [Armstrong], but it was a heck of a big leap for me,' joked the astronaut as he performed

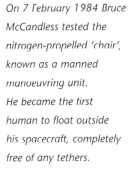

*On 7 February 1984 Bruce McCandless tested the nitrogen-propelled 'chair', known as a manned manoeuvring unit. He became the first human to float outside his spacecraft, completely free of any tethers.*

a series of gymnastic fly-pasts. Watched by a live TV audience around the world, McCandless reversed at a steady 2 mph until he was a small, white figure hovering above the blue planet. A helmet-mounted camera sent back spectacular pictures of the shuttle as it passed over the southern United States.

Despite some 'shudders and rattles and shakes', McCandless was happy to carry out a second run, this time drifting some 320 feet from the safety of the orbiter. The exercise ended with some trial runs at docking with a pin mounted on the payload bay wall.

Over the next few years, the MMU was used in some of the most spectacular EVAs of the entire programme. Only two months after its initiation, George 'Pinky' Nelson tried unsuccessfully to dock with the Solar Max satellite and bring it into *Challenger*'s payload bay. After three failed attempts

*Astronaut Dale Gardner uses the MMU to fly across to the ailing Westar satellite. By inserting a 'stinger' into the end of the satellite, he was able to move it to the shuttle cargo bay to be brought back to Earth for refurbishment.*

to clamp the docking attachment around the pin on the spinning satellite, Nelson tried to grab hold of it with his hands, treating viewers on Earth to the amazing sight of an astronaut clinging onto a solar panel in a vain attempt to stop the cartwheeling of the $2^1/_2$-ton satellite.

Later that year, Joseph Allen completed the first satellite rescue when he donned his MMU and drifted over to the stranded Palapa communications satellite. Inserting a 'stinger' (a device shaped like a skeleton umbrella) into the satellite's engine nozzle, he was able to stabilize its motion, allowing capture by the robot arm. Two days later, on 14 November 1984, it was Dale Gardner's turn to retrieve the ailing Westar satellite. As a result of their efforts, both satellites, previously thought to be insurance write-offs, were brought back to Earth and refurbished.

STS-51A was the last mission to use the MMU. Following the *Challenger* disaster, the shuttle was used for less adventurous purposes. The space chair, seen as expensive to use and maintain, was put in mothballs. Its role in assisting spacewalking astronauts was eventually replaced by the simplified aid for extra-vehicular activity (SAFER), which first showed its paces during a test flight on STS-64 in September 1994. The new model was designed as a safety back-up, bearing in mind the hectic pace of spacewalks then envisaged for the ISS.

A near duplicate version of the MMU, known as Icarus, was developed in the Soviet Union. The success of the American model was probably one of the incentives for interest in such an EVA aid. The version flown on *Mir* relied on 32 thrusters supplied by air from two 50-pint high-pressure tanks. On 1 February 1990 Alexander Serebrov carried out the first trials in open space, drifting to a distance of about 100 feet from the hatch, even though he was hampered by the need for a safety line. Four days later, it was the turn of Alexander Viktorenko.

The Icarus has never been used since. *Mir* managers preferred to use the Strela crane to move their spacewalkers around the station. It was stored in the airlock on *Kvant 2* where cosmonauts trying to exit in their cumbersome suits complained that it was getting in their way. Eventually, on 8 February 1996, Yuri Gidzenko and ESA astronaut Thomas Reiter moved it to its final resting place on *Kvant 2*'s exterior.

# 7

# ISLANDS IN THE SKY

Mankind's first tentative steps into the cosmos were made inside tiny metal canisters, with little room even to stretch the legs – Michael Collins described his couch inside the Gemini craft as no larger than the front seat of a Volkswagen. As a result, the first astronauts and cosmonauts were all under 6 feet tall, so that they could be shoe-horned into the capsules. Life in such cramped, uncomfortable conditions was just about bearable for a couple of weeks, but anything longer was out of the question. If humans were to experience zero gravity for any length of time, they would have to be given an orbital home – a space station.

## THE FIRST STEPS

Film audiences of the 1960s were struck by the beauty and symmetry of the wheel-shaped space station in *2001: A Space Odyssey*. As the structure slowly turned, it generated sufficient artificial gravity for its occupants to walk and live in a fairly normal manner. But this awe-inspiring circular structure exists only on celluloid. All the space stations so far orbited have been cylindrical monsters, disfigured by protruding antennae, solar panels and other assorted fixtures.

It is surprising that almost everything we know about space stations derives from work carried out in the former Soviet Union. Until the summer of 1998, 10 space stations had been placed in orbit, only one of which was built by a Western nation. Among these, *Mir* has been permanently occupied for most of its 13 years in orbit. Another two Soviet stations were destroyed in launch accidents before they could leave the atmosphere.

The era of space stations began with the launch of *Salyut 1* on 19 April 1971. Altogether, seven such craft bore the name Salyut in honour of the dead hero, Gagarin, but as was so often the case during the Cold War, the name was simply a Soviet attempt at misinformation. There were, in fact, two competing design bureaux, each with its own idea of what a space station should look like.

The civilian stations were designed and built by Sergei Korolev's OKB-1 bureau, later renamed NPO Energia. Their military counterparts, known internally as Almaz, were the product of Korolev's great rival, Vladimir Chelomei.

Although Chelomei's OKB-52 bureau began work as early as 1964 on the design of a station and a supply vessel, the project suffered from continual delays. In 1970 the frustrated Soviet leadership ordered the Almaz plans and hardware to be handed over to Korolev's bureau, and in little more than a year the world's first space station was ready for launch.

The first generation Soviet space stations were made up of cylindrical compartments with one docking port and a separate engine section. Electricity was provided by pairs of solar panels. They met with mixed success. *Salyut 1* remained in orbit for nearly 6 months but was occupied for only 23 days. Its first resident crew was unable to enter the station because of a problem with a docking mechanism. The second crew suffocated during the return to Earth, resulting in a long hiatus while the flawed Soyuz was modified.

There followed three flops in succession. An unnamed replica of the original Salyut was launched on 29 July 1972 but failed to reach orbit. The next attempt failed when an electrical fire led to depressurization, causing the *Salyut 2* military station to break up and plunge to Earth after just 12 days in space. The third civilian station also burned up without being occupied after a failure in its attitude control system used up all of its propellant. The Soviets tried to cover up the disaster by labelling it *Cosmos 557* rather than *Salyut 3*.

These failures came at a particularly embarrassing time, since they occurred either side of the launch of *Skylab*. However, the Soviets began to learn from their mistakes. Despite further problems with the Soyuz ferry, *Salyut 3, 4* and *5* were occupied by five crews. *Salyut 3* and *5* were primarily used for defence purposes. Their crews were all-military, they used different radio frequencies, their orbits were much lower, ideal for photographic surveillance, and they released small capsules to send back secret information. In a bizarre testimony to the prevalent Soviet Cold War paranoia, the stations were also designed to carry Nudelman rapid-fire cannon, in case they were attacked by American spaceplanes!

The first significant modification came with *Salyut 3* and *4*, which differed from the earlier stations in having solar panels that could rotate to catch as much sunlight as possible. *Salyut 6* and *7* went one step further towards a permanent human presence in space by having two docking ports, one at each end. It was now possible for two Soyuz ferries to be attached, enabling resident crews to show their replacements the ropes before handing over. Extra supplies could also be delivered by automatic Progress craft during long flights.

Despite major problems, the two stations provided invaluable experience about how to operate a space station successfully. *Salyut 6* remained in orbit for almost five years (1977–82) and was occupied by six long-term crews and 10 guest crews, a total of 27 cosmonauts, two of whom spent more than 6 months aboard.

*Salyut 7* (1982–91) also received six main crews and four short visits from guest crews. The longest occupation lasted for 237 days. *Salyut 7* remained aloft for another five years after it was abandoned in 1986. During an uncontrolled re-entry, debris from the station was scattered across the remote mountainous regions of western Argentina and Chile.

Despite the cancellation of the Almaz programme in 1979, OKB-52 continued to develop its TKS manned transportation system, complete with re-entry capsule. Although Chelomei's brainchild was never occupied by cosmonauts in the way he intended, several versions did eventually dock with the *Salyut 6* and *7* stations, bringing fresh supplies and equipment. The last of these, *Cosmos 1686*, carried science instruments instead of a re-entry capsule. It burned up with *Salyut 7* in 1991.

The TKS later formed the basis for the large modules (*Kvant 1, Kvant 2, Spektr* and *Priroda*) that make up the *Mir* station, and for *Zarya*, the first section of the new ISS.

## SKYLAB

During the 1960s, when the US government was pouring billions of dollars into its manned space programme, NASA began to put forward ambitious plans for reusable spaceplanes, orbiting space stations and manned missions to Mars.

The election of President Nixon and the decline in public support after the euphoria of the first lunar landing led to a reassessment of future prospects. NASA's first space station evolved from recycled Apollo leftovers. One of the few remaining Saturn V rockets would launch the station, while the smaller Saturn IB rockets would fly the crews to the station inside surplus Apollo command and service modules. *Skylab* itself would largely consist of an empty Saturn IVB rocket stage.

In terms of size, the 96-foot long *Skylab* dwarfed its Soviet rivals. The main workshop section alone was 48 feet long and 22 feet wide. It was so large that an internal intercom system was needed for the crew to speak to each other. However, as a design concept it was a dead end. There was no way to resupply the station. All its consumables – food, clothes, oxygen, water and fuel – were on board when it was launched. When these ran out, *Skylab* became dead in the water, an empty hulk drifting in space.

The station might never have been occupied at all without the heroic efforts of the *Skylab 2* crew. During the Saturn V launch, a protective meteoroid shield and one of the two main solar panels were ripped free by aerodynamic drag. At a single stroke, the station's power supply was halved (four smaller panels were intended only to power the solar telescope). Worse was to follow. Although the lab was successfully inserted into orbit, it became clear that the remaining solar panel had not unfolded. Then ground controllers noticed that its internal temperature was inexorably rising. Without some kind of sunshade, the station would soon become uninhabitable.

After hurried consultations, NASA engineers rigged up a makeshift parasol. Veteran astronaut Pete Conrad, with rookies Paul Weitz and Joseph Kerwin, led the rescue attempt. Their first effort to cut free the jammed solar wing met with failure, but when they entered the silent station they found its atmosphere was breathable and gas masks were not needed.

Once the parasol was deployed through a small airlock, rather like an ungainly umbrella, the internal temperature steadily dropped. However, the only way to raise power levels was to free the jammed panel. In one of the most dangerous spacewalks ever attempted, Conrad and Weitz manoeuvred a bolt cutter attachment at the end of a long pole onto an aluminium strip tangled around the panel. With Weitz heaving as if his life depended on it, the strap suddenly parted, but still the panel stubbornly refused to swing fully into position. Only by pulling on a tether attached to the panel did they persuade it to spring into position.

Despite this inauspicious start, *Skylab* was occupied by three crews. Each mission lasted significantly longer than its predecessor and set new American endurance records, which stood until the arrival of US astronauts on *Mir* more than 20 years later. The final crew, who logged 84 days, 1260 orbits and 34.5 million miles, provided the first real evidence of how people could work and survive in zero gravity. Stashed inside their command module was a treasure trove of scientific data: improved crystals and metal alloys, 20,000 pictures of Earth, 19 miles of magnetic tape and 75,000 images of the Sun, as well as some of the most detailed pictures of a comet (Kohoutek) ever obtained.

William Pogue, Gerald Carr and Edward Gibson left behind a 'time capsule' of food and drink, clothing, unused film and other items that might be of use to any future occupants. It was not to be. Five years after *Skylab* was abandoned on 8 February 1974, the station nose-dived into the atmosphere and broke apart, scattering debris over the Australian outback.

## MIR

The immensely successful *Mir* station has been in orbit since 20 February 1986. In recent years it has been the butt of much media amusement, but the station has been permanently occupied for most of its 13 years aloft, far longer than any of its predecessors and at least three times more than its design lifetime.

*Mir* followed the evolutionary chain that began with *Salyut 1*. Although the 20-ton core module is similar in size and shape to the civilian Salyuts – about 30 feet long and 13 feet wide – there are a number of significant differences. The most noticeable innovation is the six docking ports – one at each end of its main axis and four located at 90 degree intervals around a spherical transfer compartment. These enabled five additional modules to be attached to the original core section over a period of 10 years, as well as the usual traffic involving Soyuz and Progress craft. An additional docking unit on the *Kristall* module was intended for use by the Russian *Buran* shuttle, but the vehicle was scrapped before a link-up could take place. Instead, further flexibility was added in 1995 with the attachment of an American-built docking tunnel to Kristall's nose, which allowed US shuttles to dock with the station, delivering replacement crew members, food and other essential supplies.

The core module consists of two cylindrical sections, an operations area and a living area. Although there is no 'up' or 'down' in space, the cabin has been designed to ease orientation to zero G. The floor of the operations area is covered with dark green carpet, the walls are light green and the ceiling is white with fluorescent lamps. Softer pastel shades are used in the living area, but labelling and arrangement of lockers and equipment follow this bottom-to-top arrangement throughout.

The main computer access, control and navigation panels are found in the forward operations area, together with medical monitoring equipment and a bicycle exercise machine. In the centre of the living area is the galley with a table, cooking equipment and waste storage. Privacy is provided by two individual crew cabins, equipped with hinged chairs, a sleeping-bag and a porthole for admiring the view. At the rear of the crew compartment is the hygiene area with toilet, hand washing and shower (no longer used) facilities.

The rear of the core module contains the main engine and fuel tanks. Running through its centre is a 6-foot wide tunnel, which leads to the aft docking port. Since 1987 this has been occupied by the small *Kvant 1* astrophysics module. Apart from its X-ray telescopes, *Kvant 1* brought six gyros and a variety of sensors, which improve the station's attitude control. It also contains an Elektron oxygen production unit and equipment for extracting carbon dioxide and other gases from the station's air. Part of the module is pressurized to allow cosmonaut access.

The larger *Kvant 2* arrived in 1989. It consisted of three compartments, including a 3-foot wide airlock, and a central section, which could also be sealed and depressurized. Station pointing was further enhanced by an additional six gyros and 32 thrusters. A second Elektron created oxygen from recycled urine. Unfortunately, the shower cabinet did not work as well as hoped and was eventually used only for an occasional sauna.

*Kristall* was designed as an orbital factory. It carried a further two docking units, four furnaces, a biotechnology experiment, a greenhouse, an Earth observation camera and various astronomical telescopes. In practice, its furnace operations have been hampered by technical breakdowns and power shortages. It later became the entrance used by visiting shuttle crews.

*The* Mir *core module has two pressurized sections. In the foreground, cosmonaut Alexander Serebrov works in the operations area. French 'spationaut' Jean-Pierre Haignere has his hands full in the rear living area.*

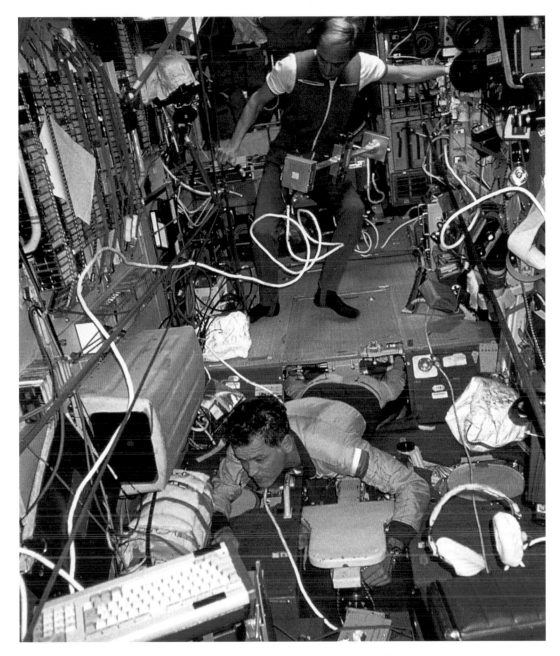

*Spektr* was added in 1995. With its four solar arrays, it became the station's main powerhouse. Its scientific equipment, mainly intended for Earth remote sensing, also included 1600 pounds of American experiments.

The station was completed in 1996 with the addition of Priroda, which, too, is used for remote sensing. A number of American experiments are also on board, including a tissue-culture area equipped with freezers and refrigerators.

## THE SHUTTLE MEETS *MIR*

The most significant single step towards cooperation in space has been the US-Russian shuttle–*Mir* programme. Not only did it lead to an exchange of astronauts and cosmonauts, but also to an exchange of ideas and greater mutual understanding between former enemies. Cultural barriers came down as suspicion dimmed and mutual respect grew.

As a result of this breakthrough, officials from East and West met on a regular basis

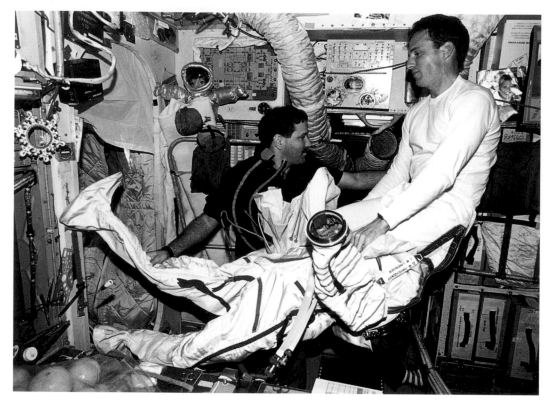

Crew change-over time on
Mir. *Astronaut Jerry
Linenger tries on his
Russian Sokol pressure suit
at the beginning of his 4-
month stay on the* Mir
*space station. Lending a
helping hand is fellow
astronaut John Grunsfeld.
On the left is a sleeping-
bag in one of the crew
cabins. The ventilation
tubes lead through to the
Kvant 1 module.*

to exchange ideas and thrash out problems. Cosmonauts and astronauts lived together, trained together, flew together, worked together and risked their lives together. Nothing could be more symbolic of the post-Cold War rapprochement than the warm embraces and sad farewells as each shuttle docked to *Mir*.

Nine times in three years, the 90-ton shuttle (usually *Atlantis*) and the 120-ton *Mir* linked up as they travelled around the planet at 5 miles a second. Their orbital ballet of rendezvous took place 2 days after launch as the shuttle took the faster, lower route to close the gap on *Mir*.

The final part of the approach took place from a point about 70 feet directly below *Mir*. After a pause for about 8 minutes of station-keeping at 30 feet to await a final 'go' for docking, the shuttle inched across the final distance to mate its docking port with the tunnel on the *Kristall* module. The firm, final embrace took place 15 minutes later as the craft completed their 'hard dock'.

Once pressure inside the two craft was equalized, the crews gathered by the tunnel to crack open the hatches and begin the welcoming ceremony. For the American visitors, the rush of air through the open hatch carried subtle reminders of *Mir*'s enduring existence. Michael Foale commented: 'It's like an old library. Except *Mir* has a very mild oily smell mixed in with the musty old books.'

The task of transferring several tons of equipment, food and water to *Mir* was deferred to the following day. The official handover between the resident American astronaut and his replacement on *Mir*, was marked by a test of the newcomer's Soyuz spacesuit and the transfer of the custom-made spacecraft seat liners for each astronaut. This body-moulded liner would be needed by the newcomer should an emergency return to Earth in a Soyuz craft be required during the coming months.

After about 5 days, the crews made their final farewells, sealed the hatches and went their separate ways.

## THE ORBITAL JUNK SHOP

After more than a decade in orbit *Mir* has gradually filled with a collection of surplus equipment, old spacesuits, spare parts and general bric-à-brac. The problem lies with the limited capacity of the Progress and Soyuz supply craft to dump or bring back to Earth any outdated or unnecessary hardware.

The contrast between the pristine, organized environment of the shuttle and the crowded, chaotic cabin of *Mir* is especially apparent to astronauts who have flown on both craft. David Wolf gave a graphic description of the problems:

*Velcro is the lifesaver for organization – but what about 150 film cans, 25 cassette tapes, 25 CDs, 40 sets of clothes, 7 cameras, 20 lenses, over 1000 components of scientific gear, 10 hard drives, 100 optical disks, 50 floppies, 2 critical PCMCIA memory cards, 4 watches, 6 computers, ... 4 months of food, 30 packs of no-rinse shampoo, 60 more of body soap, razor blades, ... and literally six tons more. The organizational/inventory task alone is daunting. Then come the radiograms instructing us to begin using all this.*

Michael Foale explained:

*There's stuff on every wall, floor and ceiling; it's all jammed in there, tied up with bungees. Once you get to the core module, there's a bit more room and you can actually do a somersault and flip – but that's the only place you can do that on* Mir.

Not only is finding enough work or leisure space a problem but actually tracking down the location of a particular piece of equipment can be a nightmare. Nothing ever seems to be stored where the documentation says it should be.

*Now where did I put that cassette? Yuri Gidzenko looks around* Mir*'s core module while Kenneth Cameron, Sergei Avdeyev and William McArthur do some sorting in the background.*

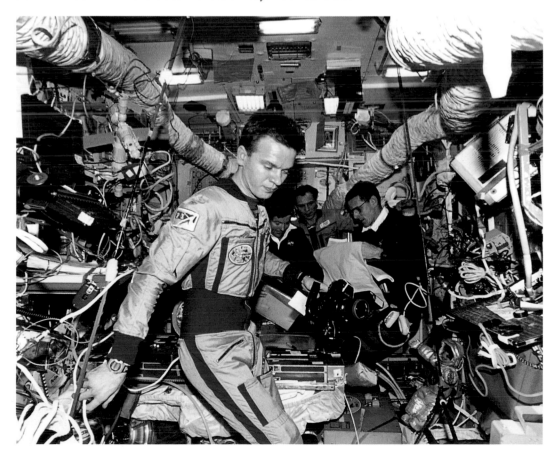

## A *Mir* Catastrophe

It was fortunate that Michael Foale could not foresee the events that would unfold during his 134-day posting when he arrived on board *Mir* on 17 May 1997. Although the station had passed its eleventh birthday and needed constant attention, the newcomer was quoted as saying it was in fine condition. Despite a fire a few months earlier and a recent breakdown in the station's primary oxygen-generation system, NASA officials were confident that Foale would be able to complete his experimental programme.

The high hopes were dashed on 25 June in an incident that almost destroyed the space station and killed its three occupants. The unmanned Progress M-34 craft, loaded with rubbish from the *Mir* station, had undocked the previous day. Now cosmonauts Vasili Tsibliev and Alexander Lazutkin were using the Progress to try out a new manual redocking system that was intended to minimize fuel consumption. If successful, the cash-strapped Russians hoped to introduce it on a regular basis. However, the omens were not good. A similar attempt the previous March had ended in failure.

Tsibliev was clearly concerned about the exercise he would have to perform. Foale explained: 'All he can judge the speed by is by how much the solar arrays are getting bigger. ... At NASA we do this with two or three different cameras in reserve, using lasers and radar.' Tsibliev's only aids would be a joystick and a TV screen.

The operation began while the crew were out of touch with ground controllers. All seemed to go smoothly at first, but concern turned abruptly to alarm when it became clear that the spacecraft was not aligned correctly and was approaching too quickly. Tsibliev tried to alter its motion and direct it away from the projecting modules and appendages of the orbital complex. Too late. The Progress swept along the side of the core module, then slammed into a solar panel and a radiator on the outer skin of the *Spektr* module, before continuing on its way. The entire station shook as the sound of the collision reverberated around the metallic modules. 'I remember thinking that I was probably going to die,' said Foale.

Almost immediately, a klaxon started sounding, warning of a hull breach. Popping in their ears confirmed the crew's worst fears: their life-giving air was leaking into space. A second alarm told the men that the station was also beginning to spin out of control, but they were forced to push this to the back of their minds. If they could not seal off the puncture in *Spektr* within 20–30 minutes they would be forced to evacuate *Mir*.

Lazutkin immediately set to work clearing the Soyuz entrance of ventilation tubes and cables. Realizing what he was trying to do, Foale joined in. They set about cutting the power cables that were draped through *Spektr*'s entrance. With minutes to spare, they succeeded in closing the hatch.

Once this was achieved, the air pressure began to stabilize, but *Mir*'s main source of power had been cut off. The crew had no choice but to shut down as many devices as possible to conserve energy, including the thermal control systems, the ventilation system, the urine processing system and the lights.

Left *A view of the* Mir *space station after the collision with a Progress spacecraft on 25 June 1997. Instead of pulling in at the rear docking port (here occupied by another Progress), the cargo craft veered left, careering into the* Spektr *module (top). The collision damaged a solar panel, caused dents in a nearby white radiator panel and punctured* Spektr, *forcing the* Mir *crew to abandon it.*

Right *Cosmonaut Valeri Korzun is surrounded by a tangle of spare parts, ventilation tubes and cables inside the* Mir *node. This central hub links the four large science modules and the Soyuz ferry to the core module. After the collision, the tubes and power lines had to be cleared from* Spektr*'s entrance before its hatch could be sealed. No longer able to sleep in* Spektr, *Michael Foale made this his new bedroom.*

Taking it in turns to work and sleep, the crew set to work to stabilize the situation. Using the Soyuz thrusters, they managed to stop the slow spin of the station caused by the collision and then aim the remaining operational solar panels towards the Sun. Foale described the scene as they strove to get the station back on its feet:

*I was 'running' from one window to the next, trying to figure out where the station was, then shouting to Vasili in the Soyuz. Then the Sun came out and it looked sort of right, so I said, 'OK, Vasili, let's go for it.'*

As the batteries charged, the men were able to bring some semblance of normality back to their lives. The lights came back on, the temperature and humidity improved, and the toilet was operational. Foale was obliged to bed down in the *Mir* node, with his feet protruding into the dark, damp *Priroda* section. His Russian colleagues lent him a toothbrush to tide him over until his lost gear could be replaced.

Just as the worst seemed to be over, the gyrodynes, which control the station's orientation system, began to malfunction.

Help finally arrived on 7 July in the form of Progress M35, which was loaded with new power cables, a hatch plate fitted with cable adapters, a shuttle medical kit, tools, computer spares and a replacement hygiene kit for Foale.

The first stage of the recovery plan was to determine the amount of damage to *Spektr* by carrying out an internal spacewalk. Then, on 15 July, came the announcement that the spacewalk was postponed indefinitely. Routine monitoring of Tsibliev's medical condition had revealed an irregular heart rhythm. One report stated that he had been affected by the 'psychological responsibility for the work that has been entrusted to him'. The exhausted cosmonaut was prescribed 'heart pills and medicines to improve his mental state and his sleep'.

Worse was to follow. On 16 July, Lazutkin, who was working solo while his colleagues slept, accidentally disconnected a cable linked to the computer that controlled *Mir*'s attitude. The others were awakened by alarms as the station drifted out of position, but it was too late. Once again plunged into darkness, the crew had to endure another major shutdown of the station while its batteries recharged.

By this time, with the men struggling to overcome fatigue and stress, mission controllers decided to entrust *Spektr* repairs to another crew. While his colleagues struggled to keep the gyrodynes spinning, Foale returned to a more relaxing occupation, planting a new generation of mustard seeds in his greenhouse.

On 7 August a Soyuz bearing spacewalk specialist Anatoli Solovyov and engineer Pavel Vinogradov pulled in at *Mir*'s rear port. A week later, Tsibliev and Lazutkin returned home, safe and well after their ordeal. Foale stayed on board to witness a gradual restoration of power levels after the installation of a new *Spektr* hatch and cable connections on 22 August. His ultimate reward for hanging in there was a 6-hour spacewalk to inspect the crumpled outer hull on 6 September.

After further frustrations involving a crashed computer and more power failures, Foale and his two Russian colleagues played host to the crew of shuttle *Atlantis*. The weary astronaut eventually returned to the comforts of home on 6 October after the second longest flight in US space history.

A number of factors were identified as having contributed to the accident: inadequate crew training, a fully loaded Progress craft that was almost a ton overweight and crew error were all mentioned.

Meanwhile, reluctant to abandon their old warhorse, Russian officials began to plan how to restore *Mir*. Reservations for paying guests were opened once more. According to current plans, the aged station's orbit will gradually be lowered until it is abandoned and allowed to burn up over the Pacific, but not before the century is out.

## ORBITAL SCIENCE

A major criticism levelled at those who favour science investigations on board shuttles or space stations is that the microgravity environment is degraded simply by having people aboard. Critics argue that a much better environment would be provided by automated, free-flying laboratories.

Continuous, barely perceptible, vibrations spread like tiny earthquakes throughout the station. Some are caused by the operation of equipment, such as fans, pumps or thruster firings. Crew members bouncing off walls or carrying out physical exercise on treadmills or bicycles also shake the station. Unfortunately, the slightest movement can affect the results of sensitive sciencific experiments and reduce the advantages of conducting them in space.

Carrying out experiments can be more time-consuming in orbit than on the ground because every piece of equipment has to be fixed in place. Viktor Savinykh gave an example of how four cosmonauts were needed to conduct a simple experiment. Their task was to photograph the layers of the atmosphere though a porthole as the Sun set on the horizon. While Savinykh and Grechko handled the photography, Vladimir Dzhanibekov tracked the Sun's movement from a second porthole, and Vladimir Vasyutin had to record the exact time of sunset. Although the experiment lasted less than a minute, it took at least an hour to prepare the crew and set up the equipment.

Support from the experts on the ground is essential if the non-specialist astronaut is to squeeze the best results from his batch of experiments.

Unfortunately, communications with *Mir* are far from reliable, particularly since the seagoing tracking ships have been withdrawn from use because of their cost. Cosmonauts can speak to Russian ground stations only around 10 times every 24 hours, each slot lasting for around 10–20 minutes. Other links using geostationary satellites are possible in

*Shannon Lucid admires wheat plants grown in the greenhouse on* Mir.

theory at any time but in reality are usually restricted to one session a day.

Apart from life sciences experiments, cosmonauts have conducted a variety of scientific investigations over the years, ranging through Earth observation, astronomy and the production of new metal alloys, superconductors and pharmaceuticals. One of the most satisfying projects has attempted to fulfil the dreams of growing fresh food in zero gravity instead of relying on rations from Earth. Unfortunately, plants find there is no obvious stimulus to direct their roots downwards and their stems upwards.

Few astronauts enjoy the more intrusive experiments such as drawing blood samples, but tending young plants or animals usually has a soothing effect, reminding the crew of the world they have left behind. As Michael Foale explained: 'I loved the greenhouse experiment. It didn't matter that the plants were tiny and minuscule.'

Until recently, plants grown in space from seeds could not be persuaded to complete their growth cycle. Foale found the answer by using a dead bee glued to the end of a toothpick to pollinate the plants. The result was the first successful completion in microgravity of the growth cycle from plant seed through maturity and back to new seed.

## LIFE ON A SPACE STATION

Cleaning, cooking, minor repairs, wiping the dishes – these are humdrum chores normally associated with everyday life here on Earth, not with astronauts.

Instead of leading exciting, romantic lives, would-be space travellers have to work hard, endure long hours training for a flight, then spend most of the time in orbit caring for experiments, working up a sweat on an exercise machine or repairing broken equipment. Andy Thomas described a normal working day on *Mir*:

*Get up at about 8.30 (Moscow time), have breakfast, start work. Around 1 or 2 I will stop and do my exercise, then clean up. We have lunch, perhaps at 3 or 4, and then we'll continue working usually to 7 or 8 in the evening, follow that with dinner, perhaps then watch a video, watch a movie, read a book, write a diary, write letters home, read e-mail ... then go to bed at about 11 or 12. This regime operates Monday to Friday. At weekends the work schedule is reduced to 3–5 hours, though there is no escape from the arduous physical fitness training.*

Despite this apparently rigid timetable, everything is a little more relaxed on a long flight. If an experiment cannot be completed one day ... well, there's always the next day or the next.

Reactions to the workload can vary. Georgi Grechko described how easy it is for a scientist-engineer to get carried away by some absorbing item of research.

*Sometimes in the morning I knew I had a very interesting scientific programme and decided not to eat all the day to save time. I would get a chocolate from the food stores to eat during the work, but at the end of the day I would find the chocolate still in my pocket. ... so much is interesting that I don't like eating or sleeping.*

Keeping the crew busy is one way of maintaining performance and alertness during a long spell in space. Andy Thomas explained:

*Each day tends to roll into the next and there comes a certain monotony and you have to use your own resources to make the life interesting, to keep your motivation going.*

Sometimes, cosmonauts become so involved in their work that they are unable to sleep. During their joint mission, Grechko and Vladimir Dzhanibekhov were snatching no more than 3–4 hours of sleep a night. On one occasion, the restless Grechko arose to carry out an experiment, but was unable to keep his eyes open. He was found fast asleep next to a porthole.

Unfortunately, there is no getting away from the everyday housekeeping. Every member of the highly trained, well-paid crews has to take his turn. Cleanliness becomes an obsession for men who would never dream of using a scrubbing brush on Earth. Inside a warm, moist, enclosed metal canister, the growth of fungi and germs is an ever constant threat, which has to be combated by regular scrubs with a bactericidal cloth. Michael Foale commented: '*Mir* has a lot of mould on the walls. It's had an awful lot of problems with dampness, tons of water actually lost on the station, and that mould causes allergic reactions sometimes ... you always carry a handkerchief.'

Anything that isn't tied down, from the smallest hair or nail clipping to the pair of metal pliers and the contents of someone's drinks bag, will drift around the cabin, posing a nuisance and possibly a threat. Jerry Linenger explained the cleanliness routine:

*We vacuum, wash down the walls with special fungus-fighting towels, and generally straighten things up every Saturday morning. Anything that flies ... usually finds its way to the intake filters of the ventilators. When vacuuming I always carry a 'lost and found' bag for all the goodies I gather. During the week I never declare anything really lost until after Saturday morning – missing pens, tools, diskettes, toothbrushes, you name it, are merely 'displaced' ... when they disappear during the week.*

Vacuuming in space can actually be a lot of fun: 'The vacuum we use resembles a small-sized, tubular-shaped one you might use – hose in front, exhaust out the back. I fly around with it tucked between my legs, exhaust aft, and it propels me along.' Other routine jobs include cleaning (and sometimes unblocking) the zero G toilet, and collecting condensation from the pipes of the air-conditioning system, a process that usually ends in the cosmonaut receiving an unscheduled shower.

Another vital daily chore is maintaining the efficient running of the station. Spare parts are sent up on each supply ship, whether they are needed or not. Michael Foale commented:

*There's a large number of systems on* Mir *that fail because of their age. They overheat and overload. Air is meant to be able to flow behind the panels of equipment to cool them down, but these airways have become blocked with debris ... . It's 'seventies technology, which uses big transistors.*

If overwork can be a problem, too much free time can also generate stresses. Foale explained: 'Some of the hardest days to actually get through were the weekends for me, days when we didn't have much to do, because that's when you start thinking about the family on Earth ... generally we were happier when we were working.'

Although the station's skin protects the crew from all but the largest space debris, it cannot afford complete protection from another health hazard, radiation. Although most harmful solar radiation is deflected by Earth's magnetic field, highly energetic particles such as cosmic rays can penetrate the station. Space station astronauts build up a hefty dose of radiation during their months aloft.

Medical doctor Valeri Poliakov, aware of the dangers, adopted a safety-first strategy during his record-breaking 14-month stay on *Mir* by sleeping in the *Kristall* module where the large batteries made of nickel and cadmium afforded him extra shielding.

*Valeri Poliakov admires the view from a porthole on the* Mir *space station during his record-breaking 14-month stay in orbit. Note the shutters on the various windows. The photo was taken from the shuttle* Discovery *on 8 January 1994.*

## WATCHING THE EARTH GO BY

Astronauts fill their rare periods of leisure in a variety of ways. Probably the favourite occupation is simply floating by a porthole and admiring the view. As they circumnavigate the world every 90 minutes, the finite size, fragility and beauty of Earth are reinforced with every passage.

The *Mir* station carries a library of books that is often extended by newcomers. Listening to music tapes or CDs with the aid of a personal stereo is a perennial favourite, and watching the latest movies on video is another traditional way of relaxing. Some crew members find that playing a musical instrument helps them to relax at the end of a tiring day, and guitars, saxophones and keyboards have all been used to soothing effect in orbit. Clearly, for such invasive activities, some degree of cooperation and awareness of the desires of the other crew members is essential. On a space station no larger than a mobile home, one person's pleasurable pastime can easily become another's nightmare.

Keeping in touch with home is vital to combat the sense of isolation and homesickness, particularly as the absence grows longer. These days, speaking on the radio has been supplemented by modern communications. David Wolf explained:

*The computer helps a lot. I cast my ballot on the computer. E-mail also goes through it and I have some CD disks, some video disks of my family ... . As the weeks go by ... it becomes more important than I imagined to keep in touch at least visually and by audio with the people you care a lot about.*

In addition, periodic press conferences and link-ups with amateur radio hams help to break the monotony and sense of isolation. Unfortunately, the Russian *Altair* relay satellite is not always in view, so opportunities for television links with *Mir* are limited to a 30-minute window.

Flight surgeons and psychologists continually monitor the crew's conversations and behaviour in order to assess how well each of them is coping with the stresses and strains. Andy Thomas described some of the ways the support team helped him to overcome these feelings:

*The NASA organizations have provided me a lot of psychological support. They've provided a nice selection of video greetings from friends and family, which I replayed the other day and to my very great surprise and enjoyment there was one there from Alan Alda, no less, who happened to be at Johnson Space Center one day and they asked him to do this and he did a wonderful piece wishing me well on* Mir.

It is this umbilical link with Earth that helps maintain sanity and prevent an overwhelming sense of isolation, although things can go wrong. Shannon Lucid was delighted when a Progress craft delivered a 'wonderful bag of new books' selected by her daughters. Having rapidly read the first book in a series, she eagerly searched for volume two, only to find it had not been packed. 'Talk about a total sense of frustration and isolation!' she wrote. Home suddenly seemed very far away.

# 8

# THE HOMECOMING

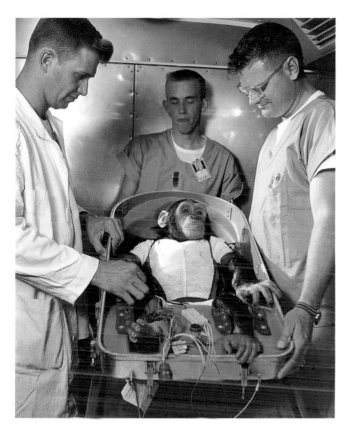

*Before the first manned missions, animals were often used to check spacecraft systems. While the Soviet Union often used dogs, chimpanzees were preferred by the United States. Here a 37-pound chimp named Ham sits in the couch that will be his home during the Mercury-Redstone 2 test flight on 31 January 1961.*

At the end of a long, exhausting spaceflight, crews begin to look forward to the comforts of home and reunions with loved ones. However, one final, forbidding obstacle remains to be overcome – the descent from orbit and the fiery re-entry through the atmosphere.

One of the worst nightmares of any space traveller is the thought of being trapped in orbit with no rescue in sight and the air supply running low. To this end, before people's lives were put at risk, the first mission planners carefully calculated orbital altitude, amounts of available oxygen and expected flight duration. These calculations were tested by precursor missions involving animals and dummies. The first problem to be overcome was to ensure that the spacecraft could be slowed so that it descended into the atmosphere. This was achieved by turning the craft so that its rear faced in the direction of movement, then firing the retro-rockets for a sufficient period to decrease velocity.

On the first missions, Soviet planners decided to rely on Nature as a back-up in case the retro-rockets failed. The Vostok craft were deliberately inserted into a relatively low orbit – 112 miles by 203 miles in Gagarin's case. This meant that, after no more than 10 days, the drag exerted by the atmosphere would slow the craft sufficiently for it to re-enter, although, of course, where the craft would eventually land remained a mystery. In the later Voskhod and Soyuz missions, a back-up retro-rocket was added in case the primary system malfunctioned.

A similar principle applied to the American Mercury missions. The capsules carried three small retro-rockets, timed to fire at 5-second intervals. Together, they would slow the capsule's speed by 500 feet per second. If only one fired, the craft would still be able to re-enter, but it would take much longer before friction with the layer of air took effect. If all three failed, the astronauts would have to rely on Nature to take its course.

## BURN UP

One of the problems of hurling a capsule into orbit at 17,500 mph is that it soaks up a huge amount of kinetic energy. This energy has to be dissipated before the capsule can return safely to Earth. The obvious way to do this is to drain it off in the form of heat. Unfortunately, too much sudden heating can incinerate any incoming spacecraft.

When a rapidly descending craft slams into the upper atmosphere, it generates a

Commander Bob Crippen monitors guidance and navigation systems during re-entry of shuttle Challenger in October 1984. The orange glow in the windows is caused by extreme frictional heating during re-entry. Pressure suits were not worn during most early shuttle missions.

shock wave, which may heat the air to as much as 3500 or 4000°C, depending on the speed of re-entry. Fortunately, most of the energy is released into the air, leaving around 1 per cent to heat and destroy the spacecraft's exterior. Nevertheless, this is quite sufficient to eat away sizeable amounts of the shell, as shown by numerous examples of charred capsules in museums around the world.

Not for the first time, the experience of the military came in useful in the search for a solution. Using knowledge gained from building missile nose cones for nuclear warheads, designers decided to coat the forward-facing surface of the manned capsules with an ablative material that evaporated slowly under intense heat. As the particles of burning matter disperse, so the heat is removed from the capsule. All the occupants experience is an orange glow, visible through the window.

One other, annoying, side-effect is a loss of communication with mission control. When the air temperature reaches its peak, between 50 and 25 miles above the ground, electrons are stripped from the atoms in the air, forming a cloud of plasma. Such ionization cannot be pierced by radio signals, and for more than 4 minutes during the most dangerous period of the re-entry ride, the crew is out of touch with the world. All the mission controllers can do is patiently wait and hope to reacquire the link.

## SEPARATION

Most manned spacecraft flown to date have had to split into pieces before the crew can return to Earth. The crew cabin is the only piece fitted with a protective heat-shield and designed for a safe re-entry. If its heat-shield is damaged, or cannot be turned forwards to face the region of maximum frictional heating, the crew are unlikely to survive the furnace of re-entry.

The first to encounter a problem of this kind was Gagarin. The instrument section attached to his Vostok refused to separate when it was time to re-enter. Under automatic control, he could do nothing but watch through his porthole as streamers of fire flared from the outer skin of his capsule until the home instrument module separated.

A similar experience befell America's first astronaut to orbit Earth. During Glenn's second orbit, ground control received a warning signal indicating that *Friendship 7*'s landing bag was deployed. This was a skirt that extended from the base of the Mercury capsule and pushed the heat-shield away from the inner bulkhead to cushion the impact during splashdown. If the heat-shield became detached from the craft before re-entry, Glenn would be incinerated like any meteor entering Earth's atmosphere. So, instead of jettisoning the retro-rockets that held the heat-shield in place, controllers decided to leave them in place as an extra insurance policy. The signal proved to be a false alarm. However, the disintegration of the retro-package presented Glenn with a tremendous firework display during re-entry.

Failure to separate the crew cabin is high on the list of a space traveller's worst fears. In the case of the Russian Soyuz, the orbiter has to break into three parts before re-entry can occur. Once the ship's aft section is pointing in the direction of movement, its main engine is fired to slow the spacecraft. It is then turned sideways, 90 degrees to the direction of flight, before the sections are separated. This ensures that the pieces will move apart, rather than collide, once they split.

Although there have been no fatalities through a malfunction in this separation sequence, there has been one near-miss. A retro-rocket failure on Soyuz TM5 occurred after the orbital module had been jettisoned, threatening to leave the two-man crew stranded in orbit without any sustenance or toilet facilities. Even worse, the command sequence began to jettison the service module, but the error was fortunately spotted by an alert commander. As a result of these near-fatal events, future missions were told not to jettison the orbital module until retro-fire had been successfully achieved.

*A Soyuz descent module touches down in a cloud of dust. The orange and white bull's-eye pattern makes the craft easier for rescue crews to spot.*

## RE-ENTRY

With the exceptions of the early Soviet Vostok and Voskhod craft, which were spherical, capsules have been designed to return to Earth with a blunt nose that comes down first.

At the beginning of a typical Soyuz descent sequence, the crew close the airtight hatches in the station and seal themselves inside the Soyuz. They don their Sokol spacesuits, close the internal hatches between modules and strap themselves into their contoured seats. Once undocked, the craft is placed in correct alignment and its retro-rocket is fired twice, once at 220 miles and again at 140 miles altitude. The Soyuz is turned sideways, ready for compartment separation at 90 miles above Earth. Some 48 minutes after the crew have entered the craft, the descent module enters the upper atmosphere at about 5 miles per second.

Coming out of the intense heating, the landing system starts operating at an altitude of 6 miles. As air pressure rises, a sensor sends a signal to blow the lid off the main parachute container. Two pilot parachutes deploy, extracting the drogue parachute so that the rate of descent slows to 300 feet per second. After 16–17 seconds, the main parachute container is released at $4^{1}/_{2}$ miles.

The heat-shield is jettisoned at $3^{1}/_{2}$ miles, shortly before full deployment of the main chute. A valve opens to equalize air pressure inside and outside the cabin. Next, a gamma-ray altimeter begins to operate, the seat shock-absorbing system is armed, and gas fills the stowage chamber so that the capsule will float if it touches down on water. At 5 feet above surface, the altimeter fires the solid fuel soft-landing engines, enabling a touchdown at 6–10 feet per second. Shock absorbers under the seats decrease landing impact. The main parachute is released to prevent dragging or tipping over in high winds.

## HITTING THE WINDOW

The angle at which the craft passes through the atmosphere is critical. If the angle is too steep, the G-loads on the crew become unbearable; if it is too flat, the craft may bounce in the atmosphere, leading to a touchdown far from the normal landing zone. The ideal compromise is a controlled descent with comparatively low deceleration loads for a crew who may have been debilitated by many months of weightlessness.

Adding wings to the spacecraft solves a lot of these problems, but it means a more complex design with extra weight and extra cost. Not surprisingly, in the rush to place a man in space, such niceties were overlooked in favour of a simple capsule returning on a shallow descent path. The spherical Vostok and blunt-nosed Mercury spacecraft were designed to generate no lift at all during re-entry. As a result, their occupants were obliged to endure deceleration forces of up to 8G, although their health was not compromised after missions lasting no more than a few days.

Things improved a little in spacecraft of the second generation. Cone- or gumdrop-shaped capsules gained some lift by being slightly heavier on one side than the other so that they would tilt slightly off-centre when air flowed past during re-entry. This meant the pilot could change the angle of descent by rolling the capsule around its longitudinal axis and pointing the so-called lift vector in a different direction. The accuracy of the landings improved considerably after this modification was introduced. Nevertheless, compared with modern aircraft, the ballistic designs gave a minimal amount of lift, barely sufficient for the job.

## EMERGENCY!

A number of emergency re-entries have taken place over the years. After the dramatic first walk in space, the *Voskhod 2* mission almost ended in tragedy when the capsule's automatic re-entry was aborted. Alexei Leonov and Pavel Belyayev were told to attempt the first manual return of the Soviet programme. The result was a fiery re-entry that incinerated the capsule's communication antennae, leaving the crew out of touch with the outside world. They crash-landed in the snow-covered Ural Mountains, more than 800 miles off course.

With no sign of habitation and darkness rapidly approaching, the cosmonauts set up a radio beacon and started a fire. Three hours later, a spotter plane found them. Supplies were dropped by helicopter, but a landing was impossible in the thick forest. The men had to endure a freezing night, huddled by their fire and listening to the howling of hungry wolves. It was hardly the triumphant return the heroes had anticipated. Not until the next day did a rescue team manage to reach them on skis. Eventually, they were taken to a clearing where helicopters whisked them back to civilization and the plaudits of the nation.

One of the most hair-raising re-entries involved the crew of *Soyuz 33*. During final approach to the *Salyut 6* space station, their craft's main engines faded halfway through an 8-second burn. Docking was cancelled and the crew were ordered to use the back-up system for an emergency return to Earth.

Even the back-up engines did not work perfectly. The motor failed to shut down

## The Triumph that Turned to Tragedy

On 24 April 1971 the crew of *Soyuz 10* pulled in at the docking port of *Salyut 1*, the world's first space station. Unfortunately, a problem with the transfer hatch meant the three men had to return to Earth without the distinction of becoming the first resident crew on an orbital station.

A second attempt was entrusted to Vladislav Volkov and two rookies, Georgi Dobrovolski and Viktor Patsayev. This time, all went well. Despite a few problems, the crew set an endurance record of 23 days aboard the station. On the evening of 29 June, the crew sealed the hatches and disengaged from the station. Dobrovolski reported that all systems on *Soyuz 11* were functioning normally. After three orbits drifting away from *Salyut*, they were ready to come home to a deserved heroes' parade in Red Square 'See you back on Earth. We are going into orientation now,' reported the commander. They were the last words heard by ground control.

The retro-rockets fired on schedule and the Soyuz modules separated. The capsule was on the way home, but a strange quiet had descended on the crew. Not even the normal telemetry registering heartbeats or breathing was being received. Only after the ship landed on the steppes of Karaganda was the terrible truth discovered. The men had suffocated on their way home. All attempts to resuscitate them were in vain. Most of the rescuers were in tears. A few hours later, it was the turn of the entire nation to mourn.

An inquiry determined that the men had died because of a rapid drop of air pressure in the cabin. Their loss was due to a simple design error. The jolt caused by the explosive separation of the *Soyuz* modules caused a pressure equalization valve to open. Normally this would operate to allow air into the cabin only when the capsule was within 3 miles of the ground. On this occasion, the air had rapidly vented into space, leaving the men gasping in a vacuum. Some reports say that Patsayev tried to close the valve manually, others that he unbuckled his seat restraint and attempted to seal the vent with his finger. Either way, he was overcome too quickly by the sudden depressurization. In just 45 seconds it was all over.

The men died because the Soviet space programme was cutting corners in its efforts to compete with the United States. In order to squeeze three men into the *Soyuz*, the crew were asked to wear only woollen flight suits and leather helmets. No allowance was made for any air loss from the cabin either on the way up or the way down.

The ashes of the dead heroes were buried in the Kremlin Wall.

after the allotted 213 seconds, so the alert commander, Nikolai Rukavishnikov, had to switch it off manually. Although the crew were on their way home, there was a downside. The reserve engines could be fired only once, whereas a normal re-entry involved two gradual deceleration burns. Rukavishnikov and Bulgarian Georgi Ivanov had to endure a vertical descent with deceleration forces building up to more than 15G. The commander later described the ballistic descent as 'like being in a blowtorch'.

Their capsule survived to make a night landing some distance from the planned landing site. The pilot of a circling rescue plane spotted the flashing beacon on top of the capsule and the cherry-red glow of the heat-shield beneath. The crew were retrieved, safe but exhausted, after their nerve-racking ride.

## EJECTION

How to return a cosmonaut or astronaut safely from a re-entry velocity of 17,500 mph to a gentle touchdown on land or sea severely tested engineers on both sides of the Iron Curtain.

Driven by a desire for secrecy, Soviet planners plumped for a landing on the wide open spaces of Kazakhstan. The major difficulty to be overcome was how to prevent the spacecraft's occupant from being crushed to death when he slammed into the solid

Earth. Designers calculated that, on a ballistic descent from orbit, even after its huge parachute billowed open, a Vostok spacecraft would descend at more than 300 feet per second and generate shock forces of around 100G on the hull as it smashed into the ground. This is, in fact, exactly what happened to *Soyuz 1* cosmonaut Vladimir Komarov when the main parachute on his craft failed to open.

Not surprisingly, the first Soviet craft were considered so unsafe for a manned landing that the occupants bailed out at 20,000 feet. Strapped into an ejector seat, when the capsule's hatch blew they were thrown clear of the rapidly descending craft and left to dangle below their orange and white parachutes before touching down on the grassy steppes.

None of this was made public at the time. The Soviet authorities insisted that Gagarin completed his historic flight inside *Vostok 1* so that his flight record was accepted by the International Aeronautical Federation. Not until years later did they admit that all of the Vostok crews ejected on the way down. These final minutes were described by Valentina Tereshkova:

*The chair was ejected with me in it. Then I separated from the chair and the parachute opened. It all happened very quickly.*

*After my parachute opened, I looked below me and saw that I was approaching a lake. We had been trained to land in water. But, of course, I did not want an unpleasant landing. The gusts of wind helped me, however, and luckily I landed not far from the lake, in a nearby field where some farmers were working. Because there was a strong wind, blowing with a speed of about 17 metres [56 feet] per second, it was not a particularly soft landing. The metal rim of my space helmet gave me a beautiful bruise to commemorate my return to Earth.*

*... I had to take off my spacesuit, open up the container which had landed near me and change into a track suit. I had to collect everything together, the spacesuit, parachute and chair and take them to the capsule about 400 metres [1300 feet] away.*

Tereshkova landed some way off-course in the Altai region. Uncertain that the authorities knew of her whereabouts, she asked the farmers to drive her to the nearest village with a telephone. The local switchboard operator must have been dumbstruck when she asked to be connected to the Kremlin. After describing her adventure to Premier Khrushchev, she went back to the field to await the rescue plane.

Clearly, it was all very well expecting a young, fit athlete to exit a capsule some 4 miles above Earth, execute a perfect parachute landing, remove a pressure suit, tidy up the site and wait to be picked up. A less bone-shaking, more reliable way to get a cosmonaut home in one piece was required.

During the Voskhod flights of 1964–5, the crew were able to complete their trip seated inside the cabin, but only by increasing the degree of risk. Since Voskhod was the same size and shape as Vostok, Korolev could squeeze several people inside only by omitting ejector seats and bulky pressure suits. Before this type of landing could be attempted, Korolev's design team had to perfect a soft-landing system. They introduced solid fuel rockets, which were automatically activated when the descent module was about 3 feet above the ground, bringing it to a virtual standstill amid a cloud of dust. Strapped into couches lined with individually shaped seat liners and protected by shock absorbers, the cosmonauts arrived on Earth, slightly shaken but safe. One cosmonaut described it as 'a soft tremor, reminiscent of the one we feel in a stopping lift'.

Occasionally this system malfunctions, giving the capsule's occupants a bone-shaking

*A sideview of a Vostok
ejector seat.*

touchdown. Such was the case with cosmonauts Vasily Tsibliev and Alexander Lazutkin, who returned in Soyuz TM-25 on 14 August 1997. 'Thank God no one was sitting in the right-hand chair, because the blow was rather strong,' was Tsibliev's comment.

Despite the presence of a number of sizeable rivers and lakes within the designated landing zone, Russian craft have regularly touched down on dry land. It was left to *Soyuz 23* cosmonauts Vyacheslav Zudov and Valeri Rozhdestvensky to unwittingly become the first (and, so far, only) Soviet crew to experience a splashdown, on 16 October 1976.

Even before lift-off things began to go wrong when the crew transfer bus broke down on the way to the launch pad. They got worse when the Soyuz rocket veered off-course, just avoided an abort and deposited them in a lower-than-planned orbit. The automatic rendezvous system failed, forcing the crew to resort to manual control. By the time they closed in on the *Salyut 5* space station, almost all their fuel was exhausted. Docking was out of the question, and there was just enough propellant for one de-orbit burn (two was the norm) and a hasty return home.

Conditions were far from ideal at the landing site and getting worse all the time. Not only was it dark but the temperature was falling and snow was being whipped up by gale force winds. As the descent module, swinging beneath its giant parachute, pierced the clouds, visibility was down to almost zero. In the white-out, none of the spotter teams in six Mil helicopters could pick up their quarry in the searchlights.

The howling wind swept *Soyuz 23* into the middle of a salt lake. As soon as the capsule hit the water, the redundant parachute began to billow in the gale, acting like a huge sail. Adrift on a choppy stretch of water and aware of the danger of drowning, the crew struggled out of their spacesuits and donned cold-weather survival gear. There was nothing else they could do but wait for rescue in the dark, clammy interior of the capsule, hoping that their metal home did not become a tomb.

It took about an hour for the helicopters to find the capsule. Frogmen succeeded in attaching a flotation collar but failed to affix a tow line. Amphibious vehicles were sent in, but they, too, floundered in the choppy waters. As the hours passed, moisture seeping into the electrical system caused an electrical short that deployed the reserve parachute. As water filled the empty housing and both chutes became waterlogged, the cabin began to settle lower. Soon the exit hatch and the radio antennae were submerged, cutting off all communication with the outside world.

Five hours after splashdown the crew were well into their emergency air supply – only two hours of oxygen remained. With ice forming on the walls of the cabin, the frozen cosmonauts hunched in their seats, trying to move as little as possible to conserve oxygen. Then, as dawn broke, one helicopter crew managed to secure a line, but, with all its waterlogged baggage, the capsule was too heavy to lift out of the lake. Instead, they opted to drag it to the shore. A lack of response from the crew caused everyone to fear the worse, but when the hatch was opened, the exhausted, shivering cosmonauts were found to be alive. In more ways than one, the sole spaceflight of Zudov and Rozhdestvensky had proved to be a bitter experience.

## SPLASHDOWN

Despite the potential for a spacecraft to sink and its occupants to drown before rescue can arrive, NASA preferred landings on water. Planners argued that the ocean would help to cushion the impact from a spacecraft that was travelling too fast. In addition, task forces of Navy ships and helicopters could be located in both the Pacific and Atlantic Oceans in case a mission had to be terminated suddenly. This capability proved useful on several occasions, notably when *Gemini 8* had to be brought back in a hurry after a sudden uncontrolled spin, and after the *Apollo 13* explosion on the way to the Moon.

Probably the splashdown that caused most concern was that of Scott Carpenter in *Mercury-Atlas 7*. After running out of fuel during re-entry, Carpenter was in no condition to control his flight path. *Mercury 7* overshot the target zone by 260 miles. For the next 50 minutes, his whereabouts and state of health remained unknown as search planes scoured the Atlantic Ocean. He was eventually found, seated in a life-raft alongside the capsule. NASA officials were not amused. Carpenter never flew in space again.

Until the arrival of the shuttle, American reliance on splash-downs was only briefly threatened during the development stage of the Gemini craft. NASA engineer Francis Rogallo experimented with the idea of using a paraglider as a landing method for space capsules at Langley Research Center, Hampton, Virginia. Although NASA lost interest in the concept, private companies picked it up, and a multi-million dollar hang-gliding industry was born.

## GLIDING TO EARTH

Today's astronauts are fortunate not to have to endure ballistic re-entries. Comfortably seated inside the shuttle, the crews experience less than 3G as they follow a tortuous glide path to Earth. Better still, they don't have far to go to be reunited with their families. With one exception, all shuttle missions have ended on the runways at either Kennedy Space Center (KSC), Florida, or Edwards Air Force Base, California.

A typical return leg begins when the engines of the orbital manoeuvring system (OMS) fire with the shuttle upside down and facing backwards over Australia. Flying under computer control, the orbiter enters the atmosphere 76 miles over Hawaii, half an hour before touchdown. After a blackout lasting about 15 minutes because of frictional heating of 1500°C, communications are restored at 34 miles altitude, 550 miles up range from KSC.

Sometimes the shuttle crosses the west coast of Mexico, then sweeps across Texas and the Gulf of Mexico. Other flights cross California or British Columbia, swing across the Midwest, and turn south over the Carolina–Georgia border. Marking its arrival in Florida with a double sonic boom, the shuttle comes in for the final approach at an angle of 22 degrees. By the time it crosses Titusville, about 5 minutes from touchdown, the shuttle is under pilot control.

Since the shuttle is a giant glider, equipped only for unpowered landings, it is essential that it hits the runway first time around. The pilots are aided on their final approach by a microwave scanning system that gives information on the range, altitude and azimuth. Additional insurance is provided by a runway that is 3 miles long and 300 feet wide, with an extra 1000 feet at both ends for emergencies. Cross-grooves are also built into the runway to prevent aquaplaning and improve skid resistance.

The wheels come down 14 seconds before the shuttle hits the runway at 215 mph. Once it rolls to a halt, assisted by air brakes and a parachute, the crew wait for ground crews to make the orbiter safe by removing gases and fluids that accumulate in unpressurized areas. Half an hour to an hour later, the doctor goes aboard for a quick check, then the crew are set free to greet their families, undergo physical examinations and speak to the press.

## LANDING PROBLEMS

Shuttle landings have been relatively trouble free, although bad weather in Florida sometimes forces a diversion to Edwards Air Force Base in California. Landing a shuttle is easier at Edwards because there is a wide choice of runways on the dry lake bed and the weather is usually easier to forecast — at KSC fog can arrive quickly, thunderstorms often build up in the afternoons and cross-winds can be a major problem. However, landing on the salt flats of Edwards is not always entirely straightforward. On 24 June 1985, one of *Discovery*'s wheels sank into the bed of the runway, which had been softened by rain. The billion-dollar orbiter had to be towed away.

In April 1985 STS-51D burst a tyre when excessive braking was used to counteract cross-winds. NASA decided to halt landings at KSC until a nose wheel steering system could be installed to avoid the need for differential braking. Although the new system was installed in time for mission 61C, bad weather and the *Challenger* disaster combined to delay the next Florida landing until STS-38 on 20 November 1990.

The fastest recorded landing was that of STS-39, which hit the Cape runway at 242 mph. Hit by a 9-knot cross-wind, *Discovery* touched down first on its right-hand wheel. Commander Michael Coats said: 'Suddenly I found myself being blown off to the left. The only way to correct that was to drop a wing to come back to the centre line. When I dropped the wing the right tyre came down first.' The fast, lopsided landing and hard application of the new carbon brakes caused shredding of one tyre and damage to all the others.

On only one occasion, in March 1982, has a shuttle touched down outside California or Florida. The third shuttle mission ended on the back-up desert runway at White Sands, New Mexico, after persistent bad weather at Edwards forced mission controllers to opt for an alternative site. The runway at Cape Canaveral was considered too dangerous for use during the shuttle's early test flights.

## A WEIGHT ON YOUR MIND

Home at last, the Russian crews open the hatch and wait for the rescue teams to arrive by helicopter at the designated landing area in Kazakhstan. Weakened by their long absence from normal gravity, they are lifted out of the capsule, guided gingerly to a chair and given a warm welcome. After a change of clothing and a brief medical examination, they are airlifted to Baikonur for further check-ups and debriefing. A long spell of rest and recuperation is the order of the day.

The delight of returning safely to Earth is often tempered by the side-effects of feeling normal gravity once again. The first to suffer severely were the two-man crew of *Soyuz 9*. Hampered by cramped conditions and limited exercise options, they were too weak to walk after their 18-day flight and had to be carried from the capsule.

Initial concerns that there might be a limit to human spaceflight endurance were later laid to rest as flights of weeks turned to months. Nevertheless, returning to 1G still brings space travellers down to Earth with a bump. After 237 days aloft, Vladimir Solovyov commented: 'I wake up in the morning and my first thought is, "Why didn't I break the bed?"' Accepting one's initial weakness comes as a surprise to people accustomed to performing effortless somersaults in orbit. John Blaha gave one example: 'I went to grab my belt buckle and my hand just slammed right back into the floor.'

Learning how to return home and live a normal life after long periods of weightlessness is one of the major challenges facing medical experts and astronauts in the years ahead. Without ways of overcoming the debilitating effects of zero gravity, mankind will never be able to move beyond Earth orbit and truly explore the endless bounds of space.

*Talgat Musabayev is lifted out of the Soyuz TM-19 descent capsule after touchdown on the steppes of Kazakhstan. He had just completed 4 months in space. The capsule was reported to have 'bounced' because of strong winds during the landing.*

# 9

# MEN ON THE MOON

It was 20 July 1969. One billion people around the globe stared at their TV screens, transfixed by a ghostly figure inching his way down a ladder, then jumping onto the ground. They were witnessing the most momentous steps in history. 'That's one small step for man, one giant leap for mankind,' declared the shimmering image. Neil Armstrong's 'step' onto the Moon had just transformed him from a test pilot and astronaut into the world's number one hero and celebrity.

Less than half an hour later it was Buzz Aldrin's turn to back out through the door on the lunar module and gingerly move down the ladder to the grey, dusty surface.

Mankind had finally left its birthplace and moved into the cosmos. For many, it was the defining moment in human evolution, the day when we ceased to be introverted and became true children of the universe. There would be no turning back from our destiny to explore and colonize other worlds.

The reality proved to be different. After the Armstrong–Aldrin spectacular, public interest dwindled and political support waned. The Nixon administration was happy to grab the glory but unwilling to pay the bills at a time when the Vietnam War was escalating and public unrest was growing. It did not take long to erase three missions from the Apollo flight manifest. By December 1972, the adventure was over. Nine Apollo spacecraft had orbited the Moon, 12 men had walked on its surface, and 842 pounds of soil and rock samples had been collected for analysis on Earth.

Although more than a quarter of a century has passed since those momentous missions, no more astronauts have disturbed the lunar dust. The human colonization of the Moon and planets seems a distant dream.

In the early 1960s everything had been so different. President Kennedy's extravagant commitment, made on 25 May 1961, to land a man on the Moon before the decade was out had provided a much-needed rallying call to a nation reeling from the fiasco of the Bay of Pigs and seemingly threatened by an increasingly confident and belligerent Soviet Union. Even the much vaunted technological prowess of the United States seemed to be in question as the Soviets notched up a string of space successes, each more staggering than the one before.

Kennedy's timing was impeccable, even if his ambition seemed foolhardy. His speech was made just three weeks after Alan Shepard had made a 15-minute sub-orbital lob into the Atlantic Ocean. In contrast, the Soviets had shaken the world with Gagarin's mission, and would soon launch Titov on a remarkable 24-hour flight around the planet.

Furthermore, the Americans were clearly lagging behind in rocket development. The Redstone did not even have the power to carry a man into orbit, while the limited lift capacity of the highly unreliable Atlas restricted the weight of the Mercury capsule to 4265 pounds, less than half the weight of the Soviet Vostok craft.

Although NASA entered the Moon Race lagging well behind, within 4 years it had overtaken its Soviet rival. This remarkable transformation came about through the greatest peacetime mobilization of industry in history. Given little more than 8 years to fulfil the dream, America's industrial giants were challenged to design and build the required revolutionary new hardware. In just 5 years (1960–65) the number of people working on the space programme soared from fewer than 50,000 to around 410,000. About 20,000 companies, large and small, contributed to the mammoth enterprise. The American taxpayers' contribution was around $25 billion. Costly it may have been, but Apollo also marked the high point in national morale and international prestige, a zenith that has never since been surpassed.

## Travelling to the Moon in Stages

At the time of Kennedy's announcement, no one had a definitive strategy for achieving his target. Until the final mission plan was thrashed out, it would be impossible to ask contractors to produce blueprints for the necessary hardware.

    Several alternative ideas circulated the NASA corridors. One possibility was to build a giant rocket called Nova, which would be capable of carrying the crew and their ship

*The* Apollo 16 *command and service module above the far side of the Moon, as seen from the departing lunar module.*

directly to the Moon. Another was to assemble a Moon craft in Earth orbit. The option finally chosen was called lunar orbital rendezvous.

The voyage would begin with the launch of a Saturn V, the most powerful launch vehicle ever built. Perched on top of this monumental structure would be two spaceships. The command and service module (CSM) would transport three men to and from lunar orbit. The crew would live inside the cone-shaped command module, while the main supplies, power and propulsion systems were located in a separate cylindrical section, the service module. These would remain attached until the final stages of the mission, when the command module would separate and plunge into Earth's atmosphere, protected by a heat-shield. The mission would end with the capsule parachuting into the ocean.

The lunar module (officially called the lunar excursion module or LEM) would carry two men onto the lunar surface, act as their home for a few days, then return them to the command module for the journey home. Since it would never fly through air, there were few constraints on its shape and strength. Grumman Aircraft Corporation came up with a strange angular creation, which measured 22 feet 11 inches high and 31 feet across with its legs extended. With a full fuel load and crew of two, it weighed in at just over 15 tons.

The LEM was divided into two sections. The lower part was the descent stage, essentially an octagonal landing platform with four legs, fuel tanks and a rocket motor to ensure a soft, accurate landing. Inside the casing were essential water and oxygen supplies, as well as the packages of scientific experiments for deployment on the surface. On later missions, a small battery-powered car, known as the lunar roving vehicle, was also stored in this section.

On top was the ascent stage with its crew cabin and another rocket engine used for the short trip back to the command module. This upper stage contained its own environmental control systems, attitude control thrusters, communications, navigation equipment and storage bays. Dual controls and displays were provided as a safety measure, so that if one man became incapacitated, the other could fly the craft back to the safe haven of the command module.

On the other hand, there was only one descent engine and one ascent engine. A failure in either of these could condemn the LEM's occupants to a long, lingering death in lunar orbit or on the surface of the Moon.

## FLY ME TO THE MOON

During the Apollo missions, Cape Kennedy (now returned to its original name, Cape Canaveral) became Moonport USA. From there, the monumental Saturn V lifted off with a three-man crew strapped inside the command module, more than 300 feet above the Florida swamps. By the time they reached Earth orbit 8 minutes later, two stages had been jettisoned. It was left to the third stage, the S-IVB, to boost them on their way across the 250,000-mile gulf to their quarry, a manoeuvre known as translunar injection.

Once the Earth began to recede into the distance, it was time to separate the CSM from the rocket. Swinging through a 180-degree turn, the ship returned to the S-IVB to extract the LEM from its housing.

Even this straightforward manoeuvre sometimes caused problems. *Apollo 14*'s docking mechanism refused to operate, forcing an exasperated Alan Shepard to resort to brute force and drive the probe into its housing at full speed.

Safely docked, the two craft pulled away from the spent upper stage and set course for the lunar seas. Three days later, they crossed into the region of space dominated by

the Moon's gravity and began to be pulled in towards it. On the early missions, safety was a prime concern, so the CSM followed a free-return trajectory. If, for any reason, the mission had to be aborted, the spacecraft would simply swing around the far side of the Moon and carry the crew back to Earth. The value of such caution was revealed during the flight of *Apollo 13*. With only the LEM's descent engine between them and catastrophe, the men's lives were saved by the option of a rapid return home.

On the last three Apollo flights, the landing sites were some distance from the lunar equator. To set the LEM down in these more interesting locations required some complex and lengthy orbital manoeuvres and a more risky flight path, which deprived the craft of a free return.

No lunar landing would have been possible without a way to allow the Moon's weak gravity to capture the rapidly moving CSM. This was achieved by firing the main engine for about 6 minutes, thereby lopping 2000 mph off its orbital speed. Unfortunately, the main engine burns during Apollo missions always took place while the crew were behind the Moon and out of touch with mission control. As a result, there were few controllers with unchewed finger nails by the time the programme was over.

Once the first, elliptical orbit was circularized at 62–75 miles above the surface by a further engine burn, the men could look ahead to climbing into the lunar module for the descent. Even then, there was plenty of scope for things to go wrong.

*Charles (Pete) Conrad inspects the* Surveyor 3 *spacecraft. Apollo 12 landed on the edge of a large crater, within spacewalking distance of the unmanned craft.*

Docking and undocking did not always go smoothly. On *Apollo 16* engine oscillations in the LEM threatened to abort the landing before it had even begun. The crew had to spend several hours going round and round the Moon before the mission was saved by switching on the back-up guidance system.

Despite intensive lessons memorizing lunar landmarks and numerous practice runs in simulators, piloting the LEM to a safe touchdown also required all the skills of its two occupants. Armstrong and Aldrin almost had to abort and return to orbit empty handed as they hovered over a large crater and searched for a landing site.

With just 2 per cent of their fuel remaining, they succeeded in setting down on the Sea of Tranquillity. 'Contact light. OK, engine stop ... . Houston, Tranquillity Base here. The Eagle has landed.' Reflecting the relief of all those around him, Charles Duke replied: 'Roger, Tranquillity, we copy you on the ground. You've got a bunch of guys about to turn blue. We're breathing again. Thanks a lot.'

Armstrong later explained: 'The auto-targeting was taking us right into a football field-size crater, with a large number of big boulders and rocks for about 1 or 2 crater diameters around us and it required ... flying manually over the rock field to find a reasonably good area.'

On a number of subsequent occasions, crews had reason to be anxious about their landing site. Of particular concern was the LEM's angle of repose. If it set down on a steep slope or with one foot in a deep crater, the ascent stage might not be able to take off. There was even the possibility that the craft might topple over. *Apollo 12* commander Pete Conrad admitted: 'We landed at about the two o'clock position, 20 feet or less from the side of the crater ... I didn't realize we were quite that close.'

Landings were especially hairy in the rugged lunar uplands. *Apollo 15*'s LEM, *Falcon*, was more than a ton heavier than its predecessors. When the crew cut the engine 8 feet above the surface, they dropped like a stone to the ground. James Irwin later recalled:

*Man, we hit hard. Then we started pitching and rolling to the side. My first thought – and I'm sure Dave's too – was, 'Man, are we going to get into an unstable position? Are we going to have to abort?' I thought surely we had ruptured something. ... Finally the vehicle stopped rolling, and came to rest on the side of a crater, and we just held our breath for 10 seconds.*

## LIFE ON THE MOON

Pleasant as it may be to float unaided through the cosmos, zero gravity has its drawbacks, most notably the havoc it plays with our body functions. Humans find it much easier to adapt to life on our nearest natural satellite, where health hazards are fewer and astronauts can still outstrip the performances of Olympic gold medallists.

Lifting large objects and leaping great heights are easy on a small world such as the Moon, where the pull of gravity is only one-sixth that of Earth. *Apollo 14* commander Alan Shepard gave a dramatic demonstration of this when he displayed his prowess with a golf club and promptly lost his ball as it sailed into the distance.

On the other hand, some of the benefits of zero gravity disappear. The *Apollo 11* crew found that bunking down in a cramped lunar module in their pressure suits was far less comfortable than floating in orbit. Aldrin found a bed on the cabin floor, while Armstrong found a spot on top of the ascent engine and rigged a strap around a vertical bar so that it formed a hammock for his feet.

Temperature control was also a problem. Turning down the liquid cooling system in the suits had little effect. 'After about three hours it became unbearable,' said Aldrin. 'We could have raised the window shades and let the light in to warm us, but that would have destroyed any remaining possibility of sleep.'

Things improved on later missions when the crews were able to sling hammocks across the cabin. The *Apollo 15* crew were allowed to remove their pressure suits and sleep in the nude. Apart from being liberated from the discomforts of cooling systems, metal rings and tight joints, the men were also free to visit the toilet without having to urinate in their disposable nappies. The result was 'the best night's sleep of the entire mission', according to Jim Irwin.

## MAGNIFICENT DESOLATION

The Apollo crews were all struck by the tremendous contrast between our green, well-watered planet and the dull, dreary wasteland that surrounded them. From Tranquillity Base, Armstrong described the scene before him:

*Out of the window is a relatively level plain created with a fairly large number of craters of the 5 to 50 feet variety and some ridges, small, 20 or 30 feet high, I would guess, and literally thousands of little 1- and 2-foot craters around the area. We see some angular blocks out several hundred feet in front of us that are probably two feet in size ... . There is a hill in view ... difficult to estimate but might be half a mile or a mile.*

*Neil Armstrong demonstrates the confined space inside a lunar module simulator. The two-man crew wore pressure suits as they flew the craft in an upright position.*

Astronauts on later missions were treated to more varied terrain. The *Apollo 15* crew were sent to explore a giant dry valley known as Hadley Rille, which is surrounded on three sides by mountains rising to 15,000 feet above a heavily cratered plain. The landscape surprised David Scott: 'All of the features around here are very smooth. The tops of the mountains are rounded off. There are no jagged peaks or large boulders anywhere.'

Large boulders were certainly in evidence at *Apollo 17*'s landing site in the Taurus-Littrow valley. Geologist Harrison Schmitt, the only scientist to visit the Moon, had a field day taking photographs and collecting rock samples. He became particularly excited by the unique discovery of orange soil. 'I gotta dig a trench, Houston. Zap me with a little cold water. Fantastic, sports fans. It's trench time.'

To everyone's surprise, a surfeit of dust turned out to be one of the great problems. On carefully studying the position of the LEM's feet on the Sea of Tranquillity, Neil Armstrong commented, 'It's very surprising, the surprising lack of penetration of all four of the foot pads ... . On descent both of us remarked that we could see a very large amount of very fine dust particles moving about.'

On the other hand, as they cavorted around the lunar plains, the Apollo crewmen reported footprints almost an inch deep in the fine-grained, cohesive coating. In no time

NASA was concerned
about the mobility of
astronauts on the Moon,
and various studies were
made to learn more about
the problems of walking in
one-sixth gravity. This
system of sling supports
was used at Langley
Research Center.

NASA was concerned about the mobility of astronauts on the Moon, and various studies were made to learn more about the problems of walking in one-sixth gravity. This system of sling supports was used at Langley Research Center.

the astronauts found their white pressure suits turning grey as fine particles adhered to them. Cleaning their suits and boots after several hours outside the LEM became a major headache. Sometimes, it even crept inside their suits and proved exceedingly irritating and difficult to remove. Another surprise awaited the *Apollo 11* crew when they removed their helmets inside the LEM. According to Aldrin: 'There was a distinct smell to the lunar material pungent, like gunpowder or spent cap pistols.'

## WALKING ON THE MOON

Making the final 'giant leap for mankind' proved to be a laborious, time-consuming process. Once the men had attached and checked the portable backpacks containing their EVA life-support systems, they had to vent all the air from the cabin before they could open the door.

Negotiating the narrow exit while backing out onto the porch also required some delicate instructions from one's partner. There were then a series of steps and a final drop onto the pristine, powdery surface. The Apollo astronauts found that the terrain of Earth's natural satellite posed a series of challenges all its own. Obviously, since there is no atmosphere on the Moon, EVA suits had to be worn outside the lunar module's cabin. However, unlike open space, there is some noticeable gravity and there is a solid surface on which to walk.

Getting around proved to be easier than anticipated, but Buzz Aldrin commented: 'You do have to be rather careful to keep track of where your centre of mass is.

Sometimes it takes about two or three paces to make sure you've got your feet underneath you.' Aldrin went on to explain some peculiarities of Moonwalking:

*It's difficult to know when you are leaning forward or backward and to what degree. This fact, coupled with the rather limited field of vision from our helmets, made local features of the Moon appear to change slope, depending on which way you were looking and how you were standing. The weight of the backpack tends to pull you backward, and you must consciously lean forward just a little to compensate. I believe someone has described the posture as 'tired ape' – almost erect but slumped forward a little … . It felt as if you could lean farther in any direction, without losing your balance, than on Earth.*

Maintaining balance while walking proved to be more difficult than the astronauts expected, mainly because the powdery dust caused their boots to slip if they tried to travel too quickly. A seemingly innocuous tumble could spell disaster for an astronaut if his pressure suit was damaged. On a number of occasions, Moonwalkers lost their footing and stumbled ungracefully to their knees in front of the watching TV millions, but, fortunately, no lasting damage was done.

Coming to a halt also presented unforeseen difficulties. 'Earthbound, I would have stopped my run in just one step – an abrupt halt. I immediately sensed that if I did that on the Moon, I'd be face down in the lunar dust,' commented Aldrin.

Although walking and jumping several feet off the ground was relatively easy, other forms of mobility proved remarkably difficult. Medieval knights in armour would have appreciated the problems faced by astronauts trying to touch their toes. Since bending the body at the waist was almost impossible, a slight flexing of the knees had to suffice.

## WORKING ON THE MOON

From a scientific point of view, the Apollo missions barely scratched the surface of lunar history and geology. However, the experiments the crews deployed, together with the soil and rock samples they brought back transformed our understanding of the solar system. By dating the samples, it was possible to establish that the Moon, Earth and other planets were born in a shooting gallery some 4.5 billion years ago, and that the Moon was probably created from one of these enormous impacts.

Laser reflectors left at the landing sites enabled astronomers to refine the Moon-Earth distance with unprecedented accuracy, while seismometers left at most of the sites were able to measure vibrations from Moonquakes and meteorite impacts. From *Apollo 12* onwards, the astronauts deployed a set of nuclear-powered experiments that could operate unattended long after they abandoned the sites. More wide-ranging surveys were conducted from orbit by the lone astronaut inside the command module.

The Moon turned out to be made of ancient lava and pulverized rock. Beneath its dusty covering, the subsurface was found to be remarkably solid, a fact that threatened to cause the landing teams a great deal of embarrassment. Each crew was expected to salute the Stars and Stripes after planting the flag in the ground. However, the apparently simple task of pushing a flag pole into loose 'soil' sometimes turned into a two-man operation. Aldrin commented, 'After much struggling, we finally coaxed it to remain upright, but in a most precarious position. I dreaded the possibility of the American flag collapsing into the lunar dust in front of the television camera.' Fortunately, the flag remained steadfastly upright. It even withstood the blast from *Eagle*'s ascent engine when the crew set off for home.

Preventing it from drooping was another matter. In the airless lunar environment, NASA had to design a telescopic arm that could click into a horizontal position and extend sideways, allowing the flag to unfurl. On the first mission, despite the crew's best efforts, the arm refused to extend fully. As Aldrin put it, 'Thus, the flag, which should have been flat, had its own permanent wave.'

Taking soil samples was a particular headache. Aldrin commented: 'The most difficult task I performed on the surface was driving those core samplers into the ground to get little tubes of lunar material for study. There was a significant and surprising resistance just a few inches down.'

Adjusting to lighting conditions was also a challenge. The Sun was a dazzling object, although it was fairly low in the black sky and cast long shadows since most of the Apollo missions arrived to avoid the heat of the lunar noon. Aldrin explained:

*The light was sometimes annoying because when it struck our helmets from a side angle it would enter the face plate and make a glare which reflected all over it. Then, when we entered a shadow, we would see reflections of our own faces in the front of the helmet and they obscured anything else that was to be seen. Once my face went into shadow, it took maybe 20 seconds before my pupils dilated out again and I could see details.*

Another strange aspect of working on the Moon was the foreshortening of the horizon. The Apollo crews struggled to gauge distances to certain landmarks, a misconception that arose because the Moon is so much smaller than Earth and its surface curvature is four times greater than on Earth.

From *Eagle*'s cabin, the horizon was only about 4 miles away. Down on the ground it appeared much closer. Neil Armstrong recalled: 'From the cockpit of the lunar module we judged our television camera to be only 50 or 60 feet away. Yet we knew we had pulled it out to the extent of a 100-foot cable.'

Misjudging the length of a cable is hardly a major handicap, but misjudging distance from base camp could have serious consequences. This proved to be a particular burden for the crew of *Apollo 14*, the last mission to explore without the aid of a rover. Tugging their tool cart along behind them, Shepard and Mitchell became disorientated. (There was no point in taking a compass along since it would be useless in the weak, localized magnetic field of the Moon.) Despite frequent map checks, they failed to find their prime target on the second EVA, the rim of Cone Crater. Only later did they discover that they had walked around the crater's lower rim without even recognizing it.

## THE LUNAR CAR

As the *Apollo 14* crew found, walking was not the ideal way to go sightseeing on the Moon. Later crews were fortunate enough to be provided with a stripped-down buggy that could carry them and all their gear up to 6 miles from the LEM. Any further was not recommended since a breakdown could leave the occupants with a long walk back to safety.

This remarkable battery-powered machine arrived on the Moon folded inside the LEM's descent stage. Yet, after simply pulling a couple of strings, removing some release pins and attaching the extra fittings, the astronauts were ready to hit the trail. Either man could drive and steer with the aid of a T-shaped hand controller. Pushing forward set the vehicle in motion; pushing sideways turned the wheels; pulling backwards applied the brakes.

On Earth a top speed of 8mph would be greeted with derision, but the astronauts were more than satisfied with its performance as they bounced and swerved around the hummocky terrain, leaving a trail of dust in their wake. 'A rock 'n' roll ride – combination of bucking bronco and a rowboat in a rough sea,' was how Jim Irwin described his first outing in the rover.

Since all three rovers took a fair pounding on their travels, the occasional breakdown was inevitable. The first rover's front steering refused to operate during the first EVA, but the problem mysteriously cleared up for the second excursion. Ironically, David Scott then decided he preferred driving with only one system after all.

The *Apollo 17* crew gave their vehicle a particularly hard time. Gene Cernan reported 'a couple of dented tyres' where the wire mesh had given way. He and Harrison Schmitt also managed to break a rear fender, which caused them to be showered in grey dust. Some 250,000 miles from the nearest garage, the men fashioned a temporary replacement by clamping spare maps onto the wheel guard.

## It's So Nice to Go Travelling, But ...

For Armstrong and Aldrin, the adventure was over almost before it had begun. Less than a day after their historic touchdown, they had to turn their minds to battening down the hatches for a reunion with Michael Collins in the command module. However, as confidence and hardware improved, so the missions gradually lengthened. The crew of *Apollo 17* spent a record 75 hours on the lunar surface and collected 249 pounds of rock samples.

As the final footprints were impressed into the lunar dust, Cernan and Schmitt unveiled a plaque on *Challenger*'s descent stage. It read: 'Here man completed his first exploration of the Moon, December 1972, A.D. May the spirit of peace in which we came be reflected in the lives of all mankind.'

The redundant descent stage, which acted as a launch platform for the upper, ascent stage, was not the only hardware left behind on the Moon. Each mission dumped more than a million dollars' worth of rubbish and equipment on the previously pristine plains, ranging from the three roving vehicles to cameras, backpacks and empty containers. Also included were some items commemorating the personal sacrifices that had made the lunar landings possible. For example, the *Apollo 11* crew left behind a shoulder badge in memory of Grissom, Chaffee and White who had died in the launch pad fire; medals honouring the dead cosmonauts Yuri Gagarin and Vladimir Komarov; and a small silicon disc inscribed with messages of goodwill from leaders of 73 nations.

Wasteful as it may seem, the removal of dead weight was essential if the LEM was ever to get off the ground. With only one small engine to kick them into orbit, there would be no second chance. Fortunately, on each occasion the ascent engine operated perfectly. Accelerating rapidly in the weak lunar gravity, the craft rose swiftly to join the command module.

All that remained was for the pilot to greet his dust-laden colleagues and help to carry their precious cargo and equipment into the cabin. Its task complete, the ascent stage was then separated and commanded to crash into the Moon, sending the seismometers jangling for hours afterwards.

During the 3-day return leg, the crew were able to relax and recount their tales. The most exciting moments came when the command module pilot was able to display his EVA prowess by collecting film and data canisters from the module's exterior.

With less than an hour to go before hitting Earth's atmosphere, the descent capsule went over to battery power as the service module was jettisoned. Travelling at velocities of 25,000 mph, the re-entry angle was particularly critical – no shallower than 5.3 degrees and no steeper than 7.7 degrees. Coming in at a shallower angle would result in the command module bouncing back off the atmosphere and becoming permanently stranded in orbit. Arriving at a steeper angle would create such high G-forces that the crew would probably be crushed to death before they hit the water.

To the crews' dismay, splashdowns sometimes proved to be quite an ordeal. On a number of occasions the command module hit the water upside down, leaving the crew dangling in their seats until the buoyancy balloons inflated. At the climax of *Apollo 15*'s odyssey, one of the three parachutes failed to open, subjecting the crew to a hefty impact as the craft smacked into the ocean.

Most frustrating of all was the period of quarantine inflicted on the crews of the early missions. Concerned about the crew carrying some lunar bug that might infect our planet, NASA officials insisted that they be kept in isolation for several weeks. Armstrong, Aldrin and Collins were immediately sprayed with disinfectant, scrubbed down with an iodine solution, dressed in rubber suits with gas masks and whisked to a converted holiday trailer known as a mobile quarantine facility (MQF). There they stayed until the MQF arrived in Houston, when they were transferred to a more comfortable lunar receiving laboratory, along with their command module and store of samples. Their incarceration ended 21 days after lift-off from the Sea of Tranquillity.

## THE SECRET RACE TO THE MOON

Despite public protestations to the contrary, the Soviet Union was also in the Moon Race. Although several years passed before the Soviets began to take President Kennedy's speech seriously, they eventually began an ambitious programme to beat their capitalist rivals.

Two separate programmes were set in motion. One was to send a two-man crew around the Moon on an *Apollo 8*-type flyby. The other was to land a lone cosmonaut on the lunar surface. Neither plan succeeded because of a combination of excessive bureaucracy, divided leadership, internal rivalries and inadequate testing.

Developmental problems with the Soyuz spacecraft and the new Proton booster combined to thwart the first plan. However, even after the success of *Apollo 8* caused the cancellation of the mission to circumnavigate the Moon, Soviet interest in the 'man on the Moon' programme continued.

The lunar landing program also ran into difficulties as the giant N-1 rocket (the Soviet equivalent of the Saturn V) repeatedly blew up soon after launch. After four failures, the new chief designer, Glushko, cancelled the programme.

Although the Soviet lunar lander flew on several unmanned Earth orbital flights, it never became operational. More than 20 years passed before the secret was revealed. There are those who believe that cancellation was the best thing that could have happened. The plan for a cosmonaut to spacewalk to and from his Moon landing craft was particularly crude when compared with Apollo practice. The Soviet lunar lander was also technologically inferior to its Apollo counterpart.

Nevertheless, 30 years on, the engines developed for the N-1 have resurfaced in the West as propulsion systems for new, commercial rockets. There is no doubt that the giant booster could have eventually flown successfully, but without a reliable lander, no cosmonaut could ever have set foot on the Moon and lived to tell the tale.

# 10
# THE ORBITAL JIGSAW

In 1984 President Reagan committed the United States to construct a huge space station, to be known as *Freedom*. Fourteen years later, after numerous debates in Congress, several major redesigns and various name changes, the first pieces of the station were placed in orbit.

One of the factors that saved the station from the political axe was the decision to bring the Russians on board. Instead of being a propaganda tool in the Cold War, the project has become a symbol of *détente*, with 16 nations contributing their technology and scientific expertise to the programme.

Trying to calculate exactly how much each participant will pay is complicated by the general preference to offer hardware or launch services instead of hard cash. However, when launch and operating costs are added, the total bill over the station's 15-year lifetime comes to between $50 billion and $100 billion. Construction and development will cost the US alone some $34 billion. Small wonder that the ISS project is described as 'the largest peacetime scientific programme of cooperation among nations in history'.

## THE INTERNATIONAL SPACE STATION

For the next decade or more, the only place to go in space will be the ISS. Laboratories, living quarters, solar panels and docking ports will be assembled, piece by piece, like some monumental abstract sculpture. This gigantic structure will be the largest, most expensive agglomeration of hardware ever placed in orbit. Pretty it won't be, but in terms of size, the ISS will stand out as the ultimate achievement (some say, folly) of the Space Age.

Its vital statistics are certainly impressive. When completed, it will weigh in at around 470 tons, almost four times heavier than *Mir*. It will be larger than two football pitches – 365 feet by 290 feet – so large that it will be hard to miss even from the ground as it sails across the night sky. In terms of volume, its six pressurized laboratories and habitation modules will add up to the equivalent of almost four *Mir* space stations.

Both of the large modules that will form the station's core are Russian-built, although the first of these, Zarya, was paid for with American cash. It was successfully launched in November 1998. Over the next 5–6 years, 45 manned and unmanned missions will be flown before the assembly sequence is complete.

The 20-ton Zarya will provide early power and propulsion for the station as well as the capability to rendezvous and dock with the Russian service module remotely. Before the service module could arrive, however, the shuttle had to deliver the US-built Node 1, a tunnel that will link Zarya to later elements of the ISS. Mission specialist Nancy Currie

*The first crew to be launched to the ISS seen during training. From left to right: flight engineer Sergei Krikalev, ISS commander William Shepherd and Soyuz commander Yuri Gidzenko.*

manoeuvred the robot arm to mate the node to Zarya, and spacewalkers Jerry Ross and James Newman connected the electrical power and data cables. Two other nodes, which are being built in Italy as part of the ESA's contribution, will be attached later during the assembly.

The second Russian section, the service module, suffered from delays because of late payments to contractors by the Russian authorities, so its scheduled launch slipped to 1999. Once it links up with Zarya, the way will be open for the first three-person crew to occupy the station.

Living accommodation on the module includes personal sleeping quarters for the crew; a toilet and hygiene facilities; a galley with a refrigerator/freezer; and a dining table. The module will have a total of 14 windows, including three 9-inch diameter windows in the forward transfer compartment for viewing docking activities; one 16-inch diameter window in the working compartment; and an individual window in each crew compartment.

Exercise equipment will include a treadmill and a stationary bicycle. The crew's waste water and condensation will be recycled to make oxygen, but it is not planned to be used as drinking water. Later on, the US part of the ISS will also be able to generate water and oxygen for the crew. An experimental system for water recycling and oxygen generation will be tested during a shuttle mission in the year 2000. The operational units should then completed and delivered to the station in 2002.

Spacewalks using Russian Orlan-M spacesuits can be performed from the service module by using the transfer compartment as an airlock. The module will also provide data, voice and television communications with mission control centres in both Moscow and Houston, Texas.

During assembly there will be a rota of three-person crews whose main jobs will be to maintain the station, assemble and activate its components. An important aspect of the policy behind selection of these crews is their familiarity with each other and their ability to understand each other verbally. The first selections include Krikalev and Dezhurov, both veterans of the *Mir* space station, who have also flown on the space shuttle, while Gidzenko, Usachev and Onufrienko have flown with US astronauts aboard *Mir* on both long-duration and shuttle–*Mir* docking missions.

Permanent human presence in space is currently set to begin with the arrival of a Soyuz TM craft early in 2000. Selection of the first crew reflected the difficulties that must be overcome when former rivals start to cooperate. Although they will be flying on a Russian spacecraft to a station that will be largely Russian-built, as the chief investor in the ISS, NASA insisted that one of its astronauts should command the historic mission to open the era of ISS occupation. Unhappy with this situation, veteran cosmonaut Anatoli Solovyov refused to take part and asked to be reassigned. So it will be the names of US Navy Captain William Shepherd (mission commander), Russian Air Force Colonel Yuri Gidzenko (Soyuz commander) and Russian flight engineer Sergei Krikalev that will enter the history books.

During their time aboard, the first solar panels will be installed and a US science laboratory should be delivered. Altogether, four shuttle flights will be needed to deliver the eight US-made solar panels. These will gradually be attached as the central truss is extended at right angles to Zarya. An extra eight, smaller panels will eventually be part of a power platform attached to the Russian research and life-support modules.

The first mission is due to last for 3 months, when the second resident crew is scheduled to arrive aboard space shuttle *Atlantis*. This time a Russian, Yuri Usachev, will be in

command. He will be joined by American astronauts James Voss and Susan Helms.

The third crew to inhabit the ISS will launch aboard a shuttle in late 2000. In command will be Kenneth Bowersox alongside Russian cosmonauts Vladimir Dezhurov and rookie flight engineer Mikhail Turin. To provide maximum flexibility in the schedule, this crew has also trained as back-up for the first resident space station crew.

The fourth resident crew will be commanded by Russian Yuri Onufrienko. Also on this 4-month mission will be Carl Walz and Daniel Bursch. They are currently scheduled to arrive at the station on board a shuttle in late 2000.

On-board living conditions will significantly improve with the arrival of the Russian service module and the US laboratory module. Other major additions will include the Japanese experiment module in 2002 and the ESA's *Columbus* orbital facility in 2003.

The last piece of the jigsaw to arrive will be the US habitation module in April 2004. The station will have a permanent crew of six or seven, three of whom will be Russians, primarily responsible for their country's modules, components and experiments. They will arrive by Soyuz and stay on board for 6 months. The other three or four crew members will usually be from the US, with occasional visitors from the other participating nations. They will spend 90 days aloft, launching and landing by shuttle. Five shuttle flights a year to the ISS are planned after completion. Four will carry supplies and equipment in pressurized logistics modules, while one will ferry non-pressurized equipment.

Not surprisingly, the number of astronauts a particular nation can have on board and the amount of laboratory time they are allowed depends on the amount each nation has invested. Since by far the largest proportion of money is being invested by the United States, it is hardly surprising that US scientists and astronauts will have more or less permanent access to the US laboratory module plus about half of the available experiment

*When it is complete, the ISS will measure 365 feet by 290 feet and have a mass of more than 400 tons. The shuttle will dock at the node attached to the centrifuge, US, ESA and Japanese laboratories.*

time in the European and Japanese labs. NASA is also entitled to about three-quarters of the non-Russian storage space, power and crew time.

Once again, arrangements are different for the Russians. Financial problems have led NASA to pay $60 million to lease part of the Russian service module during the period of construction. Otherwise, the Russians retain full control of their own parts of the station, including two research modules, which should be attached to the station in 2002–3. Crew time in these areas will be evenly divided, with half allocated to the Russians and half to the rest of the partners. However, use of Russian laboratories and facilities will be available at a price to other users. Bartering for limited time, space and use of hardware on other modules will also be an ongoing process throughout the station's existence.

Having such a large, permanent facility in orbit will expand human presence in space enormously. By mid-1998 NASA's cumulative space time for its shuttle astronauts was about 800 days, with a further 960 days or so from long-term missions to *Mir*. Assuming at least a 10-year lifetime for the new station, astronauts from all nations should clock up at least 25,000 'crew-days' aboard the ISS.

## ASSEMBLY AND MAINTENANCE

*Construction of the ISS gets under way. The Unity (Node 1) docking section (right) was linked to the Zarya module in December 1998.*

Until now NASA spacecraft have always rendezvoused and docked under human control, but the agency is developing a cheaper alternative that would fully automate such operations. A small video camera – the 'eyes' of the system – was first tested on the November 1997 STS-87 shuttle mission. Eventually, NASA will be able to perform automated rendezvous and capture with unmanned vehicles. Such operations could be used for American missions to resupply the ISS, freeing the crew from routine tasks for more important activities.

*Scenes such as this will be commonplace during assembly of the ISS. Here Steven Smith (left) and Mark Lee collect the tools they needed to service the Hubble Space Telescope.*

Nevertheless, numerous spacewalks will be essential. Between 1998 and 2003 the annual EVA time will peak at nearly 250 hours, more than four times the annual figure for shuttle spacewalks during the early 1990s.

An assessment of ISS spacewalking requirements in June 1997 indicated that a staggering 1129 hours of US spacewalk time would be needed between 1998 and 2003 – which is nearly 400 hours more than NASA's current all-time total. This includes 929 hours (77 6-hour spacewalks by two crews) for assembly, and an extra 200 hours to service and maintain the ISS. A cadre of 14 shuttle astronauts has begun intensive training in preparation for this onerous series of EVAs. One of the astronauts' most important tasks will be to assemble the robotic arm and an airlock for use during station-based spacewalks. In addition, Russian and non-US astronauts are expected to log 432 hours on assembly and another 144 hours on maintenance. This makes a total of 1705 hours in 5 years.

Even if this ambitious schedule can be maintained, assembly of such a giant structure using simple manpower would be an impossible task. To assist the spacewalkers, the ISS partners are putting together a number of versatile robotic aids to carry out the most dangerous and onerous tasks.

The first such aid to arrive will be the Mobile Servicing System, an updated version of the Canadian-built arm used on the shuttles. This will be able to move along the main truss or relocate at fixed points elsewhere on the station. Its two 'hands' will mainly be used to assemble and maintain the station, but it will also be able to manoeuvre a fully loaded shuttle. At a later date, Japanese and European-built robotic arms will be added to further assist in external experiments, maintenance and servicing.

Most of the crew-related space station assembly and maintenance tasks will involve installing and replacing externally mounted boxes, called orbital replacement units. Much of this work will be carried out from a mobile work platform attached to the end of the station's robotic arm. It will have a swivelling foot restraint to hold the feet in place while allowing the astronauts to adjust their work position. The occupants use foot pedals to move sideways through 360 degrees or roll through 180 degrees, while a knob at the base of the platform allows up or down movement of 180 degrees.

Also attached to this mobile marvel will be a tool stanchion and a grapple fixture for holding hardware during changeovers. The tool-holder will be equipped with tool boards as well as sockets to hold replacement units. Altogether, eight adjustable foot restraints, two tool stanchions and two grappling fixtures are planned for the ISS.

Apart from the robot arms and a whole range of tools, the astronauts will be able to call on two small portable cranes. These are designed to allow a single spacewalker to transport objects with a mass up to 600 pounds to sites on the station's truss. The 156-pound crane is 6 feet tall and has a boom that extends from 4 feet to 17 feet 6 inches in length. To move it, the astronaut simply turns the ratchet fittings with either a power tool or a hand crank.

Automated free-flying satellites equipped with cameras will also be available to send back pictures of inaccessible areas and inform the crew of any exterior problems.

## LIFE ON BOARD

Unlike previous space stations, there will be plenty of room for a crew of six to spread out once the ISS is completed. The crew will sleep, exercise, relax and prepare their meals in a shirt-sleeve environment where air pressure and composition are the same as that at sea level. Facilities will include shower and bathroom compartments, storage drawers and sleeping-bags.

NASA plans to make life on the ISS quite different from life on the shuttle. There will be a 40-hour working week with 2 hours of exercise. Weekends will be mostly free, with just 4 hours of routine housekeeping chores. Time for family tele-conferences will be set aside each Sunday. Each crew member will be given 8 personal holidays per year – two a quarter – to be designated before launch.

According to David Brueneman, a station planner at JSC: 'What we're going to do is just give them a list of tasks ... and then let them decide when they're going to do them.'

Microgravity science will be undertaken between the time the ISS is at maximum altitude and orbital re-boost. These will be periods of about 30 days or more when there are no vehicles docking or undocking and there are no spacewalks. Dockings will occur when ISS is in low orbit. The orbit will be raised once more by a shuttle or Progress before it departs.

Simply getting the experimental samples to the space station intact will be a major challenge. Biological material used in experiments that will be performed on board the ISS will have to be preserved to enable it to endure a journey of several days while subjected to heavy vibrations, microgravity and radiation. Scientists are looking at methods of preserving different organisms without affecting the basic properties of the specimens.

The different laboratory modules will be used for a variety of microgravity experiments. Areas of study will include:

❶  biotechnology, the study of living cells, which may lead to new drugs and vaccines;
❷  physiology, trying to discover how the human body functions;
❸  combustion, the study of burning processes;
❹  materials science, with production of new semiconductors, alloys, glasses, ceramics and metals;
❺  fluid physics, the study of how liquids behave in microgravity;
❻  investigations into how gravity affects the growth and behaviour of plants and animals.

In addition, there will be places on the truss and exterior of the modules where experiments can be exposed to the vacuum of space.

One of the main objectives will be to improve our understanding of the workings of the human body. Not only will this help astronauts to adapt better to living in space, but it should improve the chances of finding better treatments for people who suffer from physical disabilities on Earth. In the US laboratory, life sciences research will take place in racks on both sides of the wall. Here scientists will study cell and tissue growth in an effort to understand how gravity influences the ways cells join together. Scientists are particularly keen to find out how the space environment can affect humans as flights become longer. For example, cosmic radiation and microgravity may have an influence on genetic processes in biological material. Investigators are studying the effects on the DNA of fungi and plants as a step towards understanding the influence on genetic processes in general.

Other racks in the floor and ceiling will be used for studies of materials and fluids. The objective is to expand scientists' understanding of how materials form and how this process influences their properties and usefulness. It is hoped that this will lead to advances in electronics, medical instruments and metal castings. In particular, scientists will conduct experiments to find better ways of producing semiconductors and superconductors, and crystals for lasers, computer chips and solar cells.

The spread of bacteria and disease in space is a major concern. Bacteria can form a film on any surface submerged in or exposed to water, including the water systems for crew life support on board spacecraft. Experimenters will be attempting to determine the effects of spaceflight and microgravity on the formation of such biofilms.

Earth observation will also be a popular activity. Most of the populated regions of the world will be overflown by the station as it travels around the planet at an inclination of 51.6 degrees to the equator and an altitude of 220 miles.

Despite this long list of research disciplines, there are many scientists who consider the station to be an expensive white elephant. For $30 billion plus, they argue, the science community should be able to do more than study the effects of weightlessness on a few plants, animals and humans, with no guarantee of any major breakthroughs after 15–20 years of multinational effort.

## SPACE DEBRIS

No structure anywhere near the size of the ISS has ever been placed in orbit. This would not be too much of a problem if low Earth orbit simply consisted of empty space. Unfortunately, space nations have been cluttering up the near-Earth environment with all sorts of junk over the past four decades. The orbital junkyard includes active and inactive satellites as well as debris from the break-up of launch vehicles and satellites.

Today, an estimated 35 million man-made objects are orbiting the Earth, most less

*Japanese astronaut Takao Doi tries out a new crane during shuttle mission STS-87. A similar crane will be fitted to the truss of the ISS as an aid to spacewalkers.*

than $^1/_2$ inch across. However, existing military space surveillance sensors are only able routinely to observe about 8000 of the largest objects – about the size of a softball – and 80 per cent of these are in low Earth orbit, the home of the shuttle and ISS.

Debris regularly causes problems for both manned and unmanned spacecraft. Several satellites have been damaged or disabled by collisions, and the shuttle is peppered by debris on every flight. On a few occasions, windows have had to be replaced. There have also been impacts on radiators, although no leakage of coolant has yet taken place. On STS-79, for example, there was a $^1/_8$-inch diameter crater on a radiator panel.

Sometimes larger pieces of debris, rocket stages or operational satellites are picked up by tracking stations. If the approach is too close to call, the shuttle may be ordered to move out of the way. This is not easy for a large structure such as a space station. *Mir* crews are told to head for the Soyuz craft if a collision threatens or a leak occurs. In a typical incident on 15 September 1997, the station's crew had to sit for 30 minutes inside their descent module while a US military satellite whizzed past, 1500 feet from the station.

The chances of a collision involving the ISS will be increased simply because of its vast size. Its surface area will be almost 10 times greater than that of a shuttle, and the risks of being struck are increased since the station will be continuously aloft for at least 10 years. As a result, there is a 19 per cent probability of critical collisions with objects larger than about $^1/_2$ inch across during a 10-year period. NASA estimates a 5 per cent probability that such collisions would cause a catastrophic failure, resulting in the loss of a module or a crew member.

Portions of the space station have shielding to provide protection against objects smaller than $^1/_2$ inch, but shielding against larger objects is too costly. Debris from about $^1/_8$ to 8 inches in diameter is of most concern because, within this range, the debris may be too large to shield against and too small to track and avoid.

The alternative is to manoeuvre the space station to avoid those space objects that can be accurately located by the surveillance network. However, this difficult and time-consuming exercise would not be popular, since it would disrupt on-board activities, including sensitive microgravity experiments.

One partial solution would be access to more accurate information on debris heading the station's way. A recent report concluded that NASA currently has insufficient information on the size and whereabouts of this debris to predict collisions with the space station. The report went on to call for an immediate modernization of the existing surveillance radar system so that objects as small as $^1/_2$ inch across could be regularly monitored.

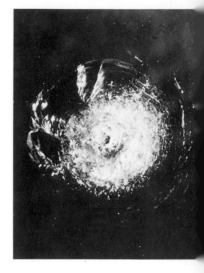

*A close-up of a mini-crater in a window caused by orbital debris during shuttle mission STS-7.*

## CREW RESCUE

As the history of the last 40 years of space exploration shows, it is as well to be prepared for the unexpected. A sudden fire or depressurization could necessitate a hurried exit for the crew. In the case of the ISS, evacuating six crew at minimal notice will not be an easy task. An interim solution will be the use of the Russian Soyuz vehicle. Two of these will be attached at all times, ready for an immediate return to Earth. But, as more people live on the station, a larger, more capable, rescue craft, able to accommodate up to seven passengers, will be needed.

NASA's answer is the X-38, a winged vehicle that uses a lifting body concept originally developed by the US Air Force in the mid-1970s. Following the jettison of a de-orbit engine module, the X-38 will execute an unpowered glide from orbit, like the space shuttle, and use a steerable parafoil for its final descent.

Drop tests are already under way, with the first unpiloted space test vehicle planned to be deployed from a shuttle for a full-scale descent and landing trial in 2000. If all goes well, the X-38 should begin operations aboard the ISS in 2003. Although the primary use of the new spacecraft would be as a 'lifeboat', the vehicle may also be modified for use as a human spacecraft.

The ESA is also intending to build an automated transfer vehicle (ATV), a small cargo craft that will be launched by an *Ariane 5* rocket to carry supplies and propellant to the ISS. It could also be available to re-boost the station's orbit and remove on-board waste as it burns up during re-entry.

Japan, too, is developing a new, unmanned space transportation system. Hope-X is a small winged shuttle, which will act as a supply ship but will also be capable of carrying out automatic landings and bringing back precious experimental products from the station. If all goes well, this should also be operational by the time the ISS is completed.

*The European Space Agency is building an automated transfer vehicle for resupplying the ISS. Set for its maiden launch in 2003, it will deliver 9 tons of equipment, water, oxygen and fuel.*

# 11
# MOON, MARS AND BEYOND

Should we be satisfied with the Apollo visits to the Moon, or should we attempt to return there, this time for good?

## WHY THE MOON?

The most obvious advantage of the Moon as a first stepping-stone into the cosmos is its relative proximity. The Apollo astronauts arrived in lunar orbit just three days after setting off from Florida. By contrast, voyages to the nearest planets would take many months, while visits to the outer solar system would take years out of an astronaut's life.

Remote sensing from orbit combined with the rocks brought back by the Apollo crews proved that the Moon is rich in some potential resources. Oxygen and silicon are common components of many rocks, while metals such as iron, aluminium and titanium are also quite abundant. These could eventually be mined and refined to make glass, ceramics and metals for construction, along with oxygen for rocket fuel and bricks made from ancient lunar lava flows. Unfortunately, most of these processes would require vast energy inputs before the resources could be separated and utilized. Although they might one day provide useful raw materials for lunar colonies or orbital bases, such lunar factories would certainly not be viable competitors for Earth-based industries.

The only lunar raw material considered to be economically useful for residents of Earth is helium-3. One million tons of this rare isotope have been deposited on the Moon by the solar wind over billions of years, and in the long term, should nuclear fusion reactors ever be developed, this reservoir of helium-3 would provide almost unlimited supplies of pollution-free energy for our energy-starved planet.

While resource exploitation of the Moon may not be possible in the near future (or even environmentally desirable), scientists are keen to promote the benefits of lunar settlement. Permanent bases could extend our knowledge of lunar geology and geochemistry, giving new insights into the early history of the Earth–Moon system and the solar system as a whole.

Astronomy in particular might benefit from a permanent lunar base. The absence of a thick, turbulent atmosphere to disturb observations means that the Moon is an ideal location to observe the universe. In particular, the lunar far side, which is always out of sight of Earth and shielded from human-generated signals, would be the ideal site to establish an observatory. The transmission of data via a satellite in lunar orbit, would enable the observatory to be largely unmanned apart from occasional maintenance visits.

Another argument favours the development of new technologies and techniques that will support mankind's eventual spread to other planets. Learning how to keep people alive through the development of artificial ecosystems, how to explore with robots, how to live off the land and how to adapt to low-gravity conditions are all seen as worthwhile goals for future lunar pioneers.

The Moon is our nearest celestial neighbour, being less than 250,000 miles away. The same side is always seen from Earth. This Apollo 17 view shows some of the dark maria on the Earth-facing hemisphere (left), but the region at right (two-thirds of this view) is normally hidden.

## THE LUNAR OBSTACLE COURSE

There are distinct advantages of life on a low gravity world compared with a weightless environment. Astronauts should be able to live more normal lives, relatively free of the damaging health effects of zero G. Settlers will be able to stride around, or brace themselves in one position while displaying the strength of an Olympic weightlifting champion.

Unfortunately, the low gravity means that there is no atmosphere to breathe. People would have to spend their entire lives either inside a pressurized habitat module or wearing a pressure suit. In addition, the lunar vacuum allows space debris, cosmic rays and charged particles from the Sun to reach the Moon's surface directly. Although they are 300 times higher than on Earth, normal lunar radiation levels should not be harmful to astronauts working outside. Large solar flares are a different proposition, but constant monitoring of activity on the Sun should allow plenty of time for astronauts to seek refuge in an underground radiation shelter.

Another problem is the huge diurnal range of temperature. In the absence of an atmosphere to transfer heat from one place to another, temperatures soar to 130°C near the equator during daylight but dip to about -170°C at night.

Long spells of darkness present another difficulty. Since the Moon rotates once every 27.3 Earth days, it follows that almost everywhere on the satellite experiences 14-day nights. Such an unusual cycle of illumination can be overcome: witness the adaptation of Eskimos and others who live in polar regions. Adaptation by plants is more of a problem. The key is a reliable energy supply, which can be harnessed to provide air-conditioning, lighting and heating.

Perpetual sunlight is available in two remote locations: the lunar poles. In these rugged spots, it is possible to construct a solar panel-laden mast on the side of a mountain or crater where the sun always shines. However, since it is much more difficult to send a spacecraft to the poles, this rugged terrain is unlikely to be settled until well into the lunar colonization programme. Elsewhere, colonies would probably have to rely on some kind of electricity generation or storage system, such as batteries, fuel cells or a nuclear power source, although more exotic alternatives are being considered.

An added incentive to locate a habitation near the lunar poles is the possibility of utilizing local water ice. Observations from the American Lunar Prospector satellite have provided evidence that a layer of ice crystals may exist in the bottom of a number of craters whose floors are permanently shaded. Estimates of the amount of water trapped in these deposits range from 11 million to 330 million tons.

Of all the resources astronauts would hope to find on another world, water must be at the top of the list. Not only can it be used for human consumption and for growing crops, it can also be split into hydrogen and oxygen. Humans require oxygen to breathe, plants need it to photosynthesize and grow. Furthermore, both gases can be used as rocket propellant.

The economics of using local resources are attractive. A relatively small amount of water would be adequate to supply 1000 two-person households for well over a century on the lunar surface, even without any recycling. On the other hand, even if future space transportation costs could be slashed to one-tenth of the current figure, it would still cost $60 trillion to lift the same amount of water into space, with further add-on costs for delivery to the lunar surface. Instead of paying vast sums to send convoys of supply ships, *in situ* processing plants could make the bases almost self-sufficient.

The main drawback is the inaccessible location of the water ice, which would require lengthy distribution networks. Much may depend on how easy it is to obtain from deep, steep-sided craters where the temperature never rises above −30°C and where the ice makes up perhaps 1 per cent of the surface material.

If lunar water does prove to be difficult or expensive to extract, various methods of recycling may be employed as partial alternatives. One of the most important areas of current research involves ways of developing a human life-support system capable of supplying food, water and oxygen, which can operate indefinitely without re-supply from Earth. Ground-based trials aimed at testing these technologies have been conducted by the Biosphere Project and by various space agencies around the world.

For example, since August 1995, NASA's Lunar–Mars Life Support Test Project has been carrying out experiments using variable pressure chambers located at the JSC. During a test which ended on 19 December 1997, the crew completed 91 days in isolation, setting the record so far for the longest duration human closed chamber test in the United States.

The next stage will be to focus on a so-called bioregenerative life-support system. This will use five chambers combining life-support technologies to provide all the air and water, and most of the food, for a crew of four. Initial testing is scheduled for 2001, culminating in a 425-day test starting in 2006.

## RETURN TO THE MOON?

Although a number of nations have shown interest in returning humans to the Moon, only the ESA and the Japanese currently have draft plans detailing how it may be achieved. In each case, however, they envisage an international effort as the only realistic strategy for the final stages.

The first step will be initial reconnaissance by a variety of unmanned orbiters, landers and rovers. These will add to our local knowledge and provide an inventory of lunar resources. Once the key technologies have been successfully tried and tested, the way will be open for a permanent robotic presence. This would include unmanned astronomical observatories, seismic stations to measure Moonquakes, and more detailed exploration using rovers equipped for tele-operated excursions.

Stage three envisages the development of automated techniques to pave the way for human settlement. These would include learning how to generate oxygen and produce construction materials, as well as the deployment of large astronomical instruments and bio-laboratories for life sciences studies.

Finally, possibly around 2015–20, the time would arrive for humans to make a permanent home on another world. Although the first homes may well be brought from Earth, the eventual aim will be to utilize local raw materials as much as possible. Once manufacturing and business firms establish themselves on the Moon, families will begin to settle, hotels will be built and the first tourists will arrive.

Feeding the community may not be too difficult as long as sufficient water is available. Using domes with an artificially controlled environment, food production should be quite efficient, albeit energy intensive. It has been suggested that vegetables raised on the Moon may well grow six times bigger than on Earth, with tomatoes and cucumbers

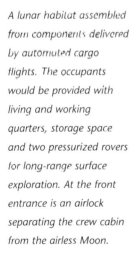

*A lunar habitat assembled from components delivered by automated cargo flights. The occupants would be provided with living and working quarters, storage space and two pressurized rovers for long-range surface exploration. At the front entrance is an airlock separating the crew cabin from the airless Moon.*

*Permanent lunar settlements will be as self-sufficient as possible. Here a 50-foot diameter inflatable habitat, covered with a thick protective outer layer, encloses a base operations centre, a small clean room, a life sciences laboratory, hydroponics gardens, a wardroom, private crew quarters, dust extractors and an airlock.*

weighing as much as large potatoes. New varieties, such as fast-growing rice that matures in 100 days, are already under development, although Japanese researchers admit they have some way to go to improve the taste.

## MOON V. MARS

An international lunar workshop held in Switzerland in 1994 concluded that the exploration of the Moon should come before the exploration of Mars. The main argument was that lunar exploration is easier and safer. Once the Moon has been successfully settled, missions to Mars will benefit from lessons learned and become more affordable. The workshop concluded: 'The biggest risks on Mars trips will be operational risks associated with long-duration missions.'

The problem is that Mars missions will last for about three years, with 12–18 months on the surface of the Red Planet. Of particular concern was the possibility of a mission failure caused by numerous malfunctions of vital hardware during such long trips. Should this happen, the crew would be isolated, unable to call for help or make an immediate return to Earth. The Moon, however, could provide a test-bed for the long-life hardware required for future Mars missions and for developing tools that would enable crews to repair and maintain their spacecraft systems.

Not everyone agrees with this approach. The 'case for Mars' supporters argue that there is no logical reason to stop off at the Moon before going on to the Red Planet. Using the Moon as a potential fuelling stop on the road to Mars is a non-starter on economic grounds. Although it may be possible to extract oxygen from lunar rocks or ice, there is no hydrogen or methane on the Moon with which to burn the oxygen. Instead of wasting time and energy diverting to the Moon to stock up with expensively produced oxygen, why not go the direct route? A Mars-bound rocket could just as easily set off from low Earth orbit as from the Moon's surface or even lunar orbit. As Mars advocate Robert Zubrin wrote: 'Basically, refuelling at the Moon on your way to Mars is about as smart as having an airplane flying from Houston to San Francisco stop over for refuelling in Saskatoon.'

As for using the Moon as a test-bed, the Mars-direct supporters believe that any remote spot on Earth, or even a dedicated space station module, would do just as well. With its lack of atmosphere, 672-hour day, extreme temperatures and entirely different landscape and geological history, the Moon can hardly be said to resemble the much larger planet.

Using reverse logic, they argue that hardware developed for a Mars mission could eventually be used to support transportation of equipment and humans to the Moon, and to establish colonies on the lunar plains.

## Why Mars?

The Moon may be a closer stepping-stone on humanity's journey to the stars, but Mars is more Earth-like and altogether more welcoming. Instead of a grey, cratered wilderness, Mars displays polar ice caps, clouds, frosts, giant rift valleys, enormous volcanoes and sinuous, river-like channels.

However, we also know that the Martian environment is far more hostile than was once believed, with air pressure less than 1 per cent of that on Earth and most of the atmosphere composed of unbreathable carbon dioxide. Anyone intending to live on Mars will have to get used to living inside a pressurized cabin and changing into a spacesuit for surface excursions.

Martian temperatures are comparable to those in Antarctica – the Viking spacecraft measured daily temperatures ranging from -30°C to -80°C. Even at the equator a thermometer would rarely struggle above freezing. Moreover, in the absence of a protective ozone layer, the surface is continually sterilized by solar ultraviolet radiation.

And yet. Images from the Mariner 9 and Viking orbiters have revealed evidence that Mars was once much more Earth-like, a wetter, warmer world where volcanoes taller than Mount Everest belched huge volumes of gas into the atmosphere, and torrents of running water carved deep, winding valleys and broad flood plains. Impact craters surrounded by splashmarks and mud flows suggest that much of this water may still exist, bonded as permafrost into the subsoil. There may even be underground reservoirs of water, heated by the remnants of that ancient vulcanicity.

Primitive life appeared on Earth within one billion years of the planet's formation. Did life also evolve during this brief, balmy period of Martian history?

Unlikely as it may be that there are micro-organisms surviving on Mars today, scientists intend to undertake a comprehensive robotic examination of the Martian atmosphere and surface terrain. During the next decade, the planet will be invaded by an armada of automated spacecraft.

Leading the way is NASA with its biennial series of Mars Surveyors. An indication of the way ahead came on 4 July 1997 when Mars Pathfinder set down at the mouth of an ancient, dried up river channel in the first demonstration of the 'cheaper, faster, better' unmanned missions advocated by NASA boss Dan Goldin. TV viewers around the world watched spellbound as a tiny rover, Sojourner, trundled down a ramp and began to crawl around the lander, investigating the numerous rocks and boulders for their chemical signatures.

Larger, more capable roving vehicles are under development. These will be deployed to scrape, dig, drill and sample the Martian surface. If all goes according to plan, rock samples should be brought back to Earth for analysis by a sample-return mission scheduled for 2005.

## THE ROAD TO MARS

The final challenge will be for the international space community to find a way to send humans across the gulf that separates our world and Mars.

Although Mars is the closest of all the planets apart from Venus, it will never be an easy place to visit. Even at the most favourable oppositions, when both Earth and Mars are on the same side of the Sun, the gap between them never closes to less than 35 million miles. During unfavourable oppositions, this gulf may widen to more than 60 million miles. Would-be human colonists must be aware that Mars is always at least 140 times further than the Moon. A one-way trip to Mars will last anything from 6 to 12 months.

There seems little doubt that a project of this size, cost and complexity, must be an international, perhaps even a global, undertaking. Crews will be drawn from a number of nations, almost certainly the countries that contribute the most resources in terms of finance and hardware. Beyond this, there will be few historical precedents to guide those who will select the astronauts. Will they prefer a single sex or mixed crew? How important will physical fitness and age be compared with a pilot's skills, scientific expertise and engineering know-how?

Mars advocate Robert Zubrin favours a crew of four, which would be large enough to carry out the mission, not too burdensome on the weight and consumables, and not too many to mourn if disaster strikes. Each should be a specialist in his or her own subject, although additional expertise in other fields would be essential. Since mission failure is most likely to be the result of a technological glitch, Zubrin argues that two mechanics – that is, flight engineers – must be included in any expedition.

Making sure that the crew reach their destination is one thing. Making the most of their time on Mars is another. With two scientists aboard – one geologist and one bio-geochemist – the crew would be equipped to study the local rocks, ice sheets and soils and also to probe deeper into the likelihood of life, past or present.

'There is no need for people whose dedicated function is "mission commander", "pilot" or "doctor",' says Zubrin. These functions can be fulfilled by any of the four crew members.

How will voyagers to Mars deal with a medical emergency millions of miles from home? Simple ailments such as stomach upset, cuts and abrasions are already handled on a routine basis by the judicious use of the on-board 'black bag' and, if necessary, expert advice from down below. For example, in May 1995 Gennadi Strekalov scratched his hand while cleaning the wall of the *Mir* station. Doctors on the ground were able to study a downlinked video of the injury and advise on the correct medication to apply.

Such incidents could easily be dealt with if at least one astronaut had a fairly advanced medical knowledge and each Mars crew member had training in first aid. Under those circumstances, even something as serious as a broken limb could probably be handled quite safely with the help of an on-board instruction manual and advice from consultants back on Earth.

More serious illness requiring hospital treatment is another matter. On several occasions, mission-threatening medical problems have arisen on long-duration flights. Most of these have involved a variable heart rhythm. In 1987, flight surgeons recorded an irregular heartbeat for first-timer Alexander Laveikin. The unfortunate cosmonaut was brought home halfway through his flight. He has been grounded ever since, even though the medical team later admitted they had made a mistake.

Vasili Tsibliev was prescribed medication, including sleeping tablets, taken off the heavy exercise regime and given more time for R & R when his heartbeat went haywire after he crashed a Progress spacecraft into the *Mir* space station in June 1997.

Shortly after beginning command of his first space mission in 1985, Vladimir Vasyutin became seriously ill and was hurriedly brought back to Earth. Rumours circulating at the time suggested that he had suffered a nervous breakdown, but according to cosmonaut Alexander Volkov:

*Vasyutin faced a whole complex of medical ailments, which were impossible to cure in space. He had to return to Earth quickly or his life would have been in danger. ... He suffered from strong internal pains and from a change of size of his internal organs. His liver became so bloated that it was visible on his body.*

There is an outside chance that an infection may be carried aboard by one crew member (although they will be carefully screened and kept in quarantine before lift-off). An example of what could happen came in 1968 when, despite all precautions, the crew of *Apollo 7* developed streaming colds, which caused a rapid deterioration in crew morale, a souring of crew–ground control relations and the threat of a mutiny.

It may even be possible for space radiation to cause mutations to occur in normally benign organisms, creating a new, and possible more virulent, form. Some evidence for this type of transformation was found in the early Soviet space programme. Cosmonaut Vitali Sevestyanov recalled:

*After Andrian Nikolayev and I returned from orbit [on 19 June 1970], they literally put us into custody. For five days they fed us through a safety, bio-interface system because it transpired that a mutation of two microbes occurring here on Earth took place in our metabolic systems. They proliferated very rapidly at first, but after five days they all became extinct under the action of gravity.*

What about psychological needs? Crew compatibility will be a major priority. Studies of people enclosed in confined spaces for long periods show that minor incidents and differences of opinion can swell and develop into situations where individuals harbour resentment, avoid contact and refuse to communicate.

On occasion, such clashes have arisen in space stations. Relations became so soured that crew members spent most of their free time alone at opposite ends of the space station. Cosmonaut Alexei Gubarev admitted that he and Georgi Grechko 'sometimes had different ideas about the same developments'. Gubarev noted that the normally cheerful, calm Grechko became nervous and irritable. 'We smoothed out our differences, forgave one another and became reconciled with each other's deviations in actions and behaviour.' So did everything return to normal? 'In many ways we succeeded,' was Gubarev's guarded comment.

Examples from Russian space stations suggest that the presence of women can have a 'civilizing' influence on male members of the crew, but would this necessarily be true for a crew trapped in a confined space for several years with no guarantee of a safe return? Sexual relationships, and possibly rivalries, might develop among the crew. This seems to have been the case when a four-man, four-woman team was locked for two years inside an ecological experiment called Biosphere-2. One member commented: 'People are people. Everything you might expect to happen with people has happened in here.'

Radiation is certainly a hazard to anyone leaving Earth's atmosphere. While there is no possibility of protecting a Mars crew from the steady rain of high-energy cosmic rays, solar flares can be predicted and protective measures taken. An essential part of any Mars craft would be a central storm shelter, which could shield the astronauts from sporadic blasts of protons.

Crews would also have to learn to cope with being isolated and far from home. Communications would become increasingly tedious as the Earth-spacecraft distance steadily increased. A $2\frac{1}{2}$-second time delay between sending a message from the Moon and receiving an answer is perfectly acceptable. Waiting for up to 40 minutes is a different proposition. There may also be constraints over what news to convey and what to conceal. Soviet ground controllers, for example, decided not to inform Georgi Grechko that his father had died until he returned to Earth.

## THE FIRST MARTIANS

The decision on how long the first explorers should stay will be based on orbital dynamics. Since oppositions occur only at intervals of around 780 days, planners will have to decide whether the crew should stay for a few weeks or settle down for a long wait before the planets align themselves once more. The more conservative approach favours an initial 30-day expedition on the surface, followed by further stays of up to 90 days, eventually culminating in a series of 600-day-long marathons.

Some preliminary studies have been made of how best to prepare for such a difficult, dangerous programme. In 1989 US President Bush's Space Exploration Initiative attempted to kick start the next stage in the human exploration of the solar system. The outcome was a report by the National Space Council that saw the establishment of a lunar outpost some time between 1999 and 2007, and the first expedition to Mars some time between 2010 and 2024.

One of the key stages in this ambitious scenario was the establishment of a lunar oxygen plant capable of manufacturing fuel for the Mars transfer vehicles. A new, heavy-lift launch vehicle would also have to be developed to carry the Mars hardware to the space station for assembly.

According to this scheme, the breakthrough would be made by a crew of four. Their craft would aerobrake into orbit, then descend to the surface for a 30-day stay. The lander would include a habitat module with an airlock and life-support systems. While the crew used unpressurized rovers to reconnoitre the locale, teleoperated vehicles would wander up to 30 miles from the base, studying geology and searching for useful resources. On their return, the crew would transfer to a separate capsule for the high speed re-entry into Earth's atmosphere.

This initial reconnaissance would be quickly followed by the launch of a cargo ship loaded with a permanent habitation facility. Once this was activated by the next crew to arrive, the way would be clear for a lengthy occupation, supported by the extraction of water from the permafrost subsoil and the production of oxygen for life support and rocket fuel.

This $450-billion effort spread over 30 years came in for a lot of criticism. As a result, studies by non-government engineers, scientists and Mars enthusiasts led to a series of 'Case for Mars' conferences, at which innovative and relatively low-cost ideas were put forward as alternatives to the 'Rolls Royce' approach.

## MARS DIRECT

One of the leading advocates of this approach has been former Lockheed Martin engineer, Robert Zubrin. He argues that, with careful planning and the utilization of Martian resources, humans could set foot on the planet for $20–30 billion. Spread over 20 years – 10 years to develop the hardware and another 10 years of flying missions – the cost should not become an insurmountable obstacle.

With this prodding, NASA also seems to have realized that its Mars exploration plans need to be more modest. The agency's latest (unfunded) proposals foresee a step-by-step approach before sending a six-person crew to spend almost two years on Mars, perhaps around the year 2019.

The adventure would begin with the launch of three unmanned craft on low-energy trajectories to Mars.

Six months later a fully fuelled Earth return vehicle (ERV) would aerobrake into Martian orbit, ready for a future crew to use on completion of their mission. The second craft would deliver a surface outpost to a pre-chosen exploration site. It would be equipped with a nuclear power plant, a propellant production unit, a supply of liquid hydrogen and a fuelled Mars ascent vehicle (MAV). Once on the surface, a small vehicle carrying a mini-nuclear reactor would trundle a short distance from the lander and begin to produce methane and oxygen. Over the next year it would automatically create 30 tons of rocket fuel from the carbon dioxide atmosphere and hydrogen feedstock.

*Mars exploration will begin with an unmanned surface outpost. This will manufacture oxygen and methane rocket fuel using hydrogen imported from Earth and carbon dioxide from the Martian atmosphere. Perched on top is the conical ascent vehicle, which will eventually use this fuel to lift the first crew off the planet.*

The third vehicle would land nearby with the crew habitat module, laboratory, non-perishable consumables and a second nuclear power plant.

Once mission controllers were able to verify that the tanks of the MAV were full, the way would be clear for the second stage to begin. During the next launch window the first crew would set off for Mars, together with a back-up ERV and surface outpost. The astronauts would take a faster route to minimize the hazards of prolonged weightlessness and radiation exposure, arriving two months before the cargo ships. The next giant leap for mankind would begin with the crewed habitat module touching down on the arid, rust-coloured plains.

This will just be the start. Each launch opportunity would see the delivery of a new habitat module to the Martian surface. These could be sent individually to an unexplored site or linked by tunnels so that one of the more suitable locations could be expanded into a large base camp. With the aid of pressurized vehicles for astronauts, balloons and robotic rovers, human explorers could spread out to reconnoitre the surrounding terrain for hundreds of miles around each homestead.

Living off the land is one of main principles behind the cheaper, faster approach to settling Mars. Just as the first pioneers opened up the American West with only the bare necessities, so the first voyagers to Mars should be prepared to generate their own water and oxygen, grow their own food, provide their own rocket fuel and, eventually, build their own houses.

Habitat modules are fine for temporary accommodation, but eventually a larger, more permanent form of housing will be needed. Once the inevitable teething problems are ironed out, the first true settlers will arrive, attracted by a demand for specialized labour and salaries many times higher than on Earth.

Encapsulated inside pressurized, transparent domes made of toughened plastic, residents will be able to grow crops and construct houses from local raw materials using relatively traditional, unsophisticated techniques.

Although water no longer exists in liquid form on Mars, large supplies of ice seem to be available in the polar caps and as subsurface permafrost. Energy generation should not be too much a problem, either. The first habitats could be powered by a small nuclear reactor, backed up by solar panels and batteries or fuel cells. Although sunlight on Mars is weaker and the air is much thinner than on Earth, the potential is there to generate solar and wind energy. Geothermal energy from hot underground springs close to the extinct volcanoes may also be available.

The production of rocket fuel from Martian air should not be a problem. Test plants run on Earth show that this is possible using relatively simple, reliable technology. Confirmation of the process is expected from a robotic Mars mission scheduled to land on Mars in 2001–3.

While it is advisable to bury any habitations on the Moon as protection against cosmic radiation, the Martian atmosphere will filter out most of the harmful micrometeorites and high-energy particles. Clusters of self-supporting domed settlements would be free to expand across the open deserts. Plants could also be grown in thin-walled, inflatable plastic domes resistant to ultraviolet radiation. These domes, each 150–300 feet wide, would be light enough to be carried from Earth, although, as industrial capacity grew, they could eventually be made from local resources. With an Earth-like day, adequate light levels, a regular supply of water and an artificially enhanced carbon dioxide atmosphere, plants should be able to thrive, although livestock are likely to be a rarity and most Martians will be obliged to be vegetarians.

Most importantly, the first settlers should be certain of one day returning to their home world. For the early missions, the Apollo approach is likely to be used. After 500 days on Mars the crew will mothball their Martian home, climb aboard a small ascent craft filled with locally refined propellants, then ascend to dock with a larger Earth return vehicle in low Mars orbit. This becomes their home for the 6-month return journey. For the final leg of the flight, a re-entry capsule will parachute or paraglide them back to worldwide rapturous acclaim.

## COLONIZING THE COSMOS

For the foreseeable future mankind's exploratory ambitions will be restricted to the inner solar system. In 100 years from now, spacecraft should be regularly shuttling visitors and cargo between Earth, the Moon, Mars, the Martian satellites Phobos and Deimos, and the asteroids. A regular supply-cum-tourist route will develop between these worlds, opening up the wonders of the solar system to anyone who can afford to pay. Spaceflight really will have become routine, much as air travel is commonplace today.

Inevitably, too, mining, manufacturing and commerce will spread beyond our small world, opening up the vast resources that lie on our celestial doorstep. Most industrial processes will be automated, with a handful of specialists overseeing upgrades and maintenance. The time will come when an entire generation will be born, live and die on a world other than Earth. Human bodies will evolve in different directions as they adapt to differing levels of gravity. Some people may never visit our world during their entire lifetimes. As heart muscles shrink and bones weaken, some will become too weak to visit the original home planet. The human race will truly have become children of the Universe, separated for ever from their cradle, the blue planet – Earth.

## Abbreviations

| | |
|---|---|
| ATV | automated transfer vehicle |
| AXAF | Advanced X-ray Astrophysics Facility |
| capcom | capsule commander |
| CNES | Centre National d'Études Spatiale |
| CRV | crew return vehicle |
| CSM | command and service module |
| CTV | crew transfer vehicle |
| DLR | Deutsches Centrum für Luft-und-Raumfahrt |
| DOD | Department of Defense |
| ERV | Earth return vehicle |
| ESA | European Space Agency |
| ESRO | European Space Research Organization |
| EURECA | European Retrievable Carrier |
| EVA | extra-vehicular activity |
| FGB | Functional Energy Block |
| HST | Hubble Space Telescope |
| ISS | International Space Station |
| JSC | Johnson Space Center |
| KSC | Kennedy Space Center |
| LEM | lunar excursion module |
| MAV | Mars ascent vehicle |
| MMU | manned manoeuvring unit |
| MOL | manned orbiting laboratory |
| MQF | mobile quarantine facility |
| NASA | National Aeronautics and Space Administration |
| NASDA | National Space Development Agency (Japan) |
| NPO | Research and Production Association (Russia) |
| OMS | orbital manoeuvring system |
| psi | pounds per square inch |
| RLV | reusable launch vehicle |
| RSC | Rocket Space Centre (Russia) |
| ROTEX | Robotic Technology Experiment |
| SAFER | simplified aid for extra-vehicular activity |
| SRB | solid rocket booster |
| SSME | space shuttle main engine |
| STS | Space Transportation System (space shuttle missions) |

# INDEX

Photographic Acknowledgements

P. Aventurier for ESA 106; Peter Bond 26, 51, 75, 103; © CNES 57, 87, 99; ESA 37; ESA/D. Ducros 127; © Finley Holiday Films 62, 65, 82; Freud Communications 50; Lockheed/NASA 60; NASA 1, 2, 7, 8, 10, 13, 14, 19, 21, 25, 27, 31, 32, 35, 38, 41, 43, 45, 48, 54, 63, 67, 71, 80, 81, 88, 89, 90, 91, 92, 95, 97, 98, 108, 110, 112, 113, 118, 120, 121, 122, 125, 126, 129, 131, 132, 137, 138; NASDA/NASA 61; © Sotheby's 9; © SYGMA/CNES 74.